Efficient Node.js
A Beyond-the-Basics Guide

Samer Buna

O'REILLY®

Efficient Node.js

by Samer Buna

Published by O'Reilly Media, Inc., 1005 Gravenstein Highway North, Sebastopol, CA 95472.

O'Reilly books may be purchased for educational, business, or sales promotional use. Online editions are also available for most titles (*http://oreilly.com*). For more information, contact our corporate/institutional sales department: 800-998-9938 or *corporate@oreilly.com*.

Acquisitions Editor: Amanda Quinn	**Indexer:** Judith McConville
Development Editor: Jeff Bleiel	**Interior Designer:** David Futato
Production Editor: Gregory Hyman	**Cover Designer:** Karen Montgomery
Copyeditor: Miah Sandvik	**Illustrator:** Kate Dullea
Proofreader: Stephanie English	

January 2025:   First Edition

Revision History for the First Edition

2025-01-08:   First Release

See *http://oreilly.com/catalog/errata.csp?isbn=9781098145194* for release details.

978-1-098-14519-4

[LSI]

Table of Contents

Preface. vii

1. Node Fundamentals. 1
   Introducing Node   1
      The JavaScript Language   3
      Executing Node Code   4
      Using Built-In Modules   7
      Using Packages   10
      ES Modules   15
      Asynchronous Operations   19
   The Non-Blocking Model   21
   Node Built-In Modules   28
   Node Packages   30
   Arguments Against Node   31
   Summary   32

2. Scripts and Modules. 35
   Node CLI   35
      Options and Arguments   36
      Environment Variables   40
      REPL Mode   43
   Node Modules   49
      Resolving Modules   49
      Loading Modules   50
      Scoping Modules   51
      Executing Modules   53
      Caching Modules   57
   Summary   58

3. Asynchrony and Events **59**

Sync Versus Async Handling   59

Handler Functions   63

    Promises   65

    async/await   67

    An Analogy for Promises   69

The Event Loop   72

Event Emitters   74

    Asynchrony   76

    Errors   78

    Examples   79

Summary   81

4. Errors and Debugging **83**

Throwing and Catching Errors   83

Types of Errors   85

    Standard Errors   85

    System Errors   87

    Custom Errors   89

Layered Error Management   91

Debugging in Node   94

Preventive Measures   96

    Code Quality Tools   96

    Immutable Objects   96

    Testing   97

    Code Reviews   97

Summary   97

5. Package Management **99**

Introducing Package Management   99

The npm Command   102

Semantic Versioning   108

Updating and Removing Packages   109

Creating and Publishing Packages   113

npm Run Scripts   115

The npx Command   117

Summary   119

6. Streams **121**

Introducing Streams   121

Using Streams   123

Fundamentals of Streams   126
The pipeline Method   127
Stream Events   129
Paused and Flowing Modes   130
Implementing Streams   131
    Writable Streams   131
    Readable Streams   132
    Duplex/Transform Streams   135
Async Generators and Iterators   137
Streams Object Mode   139
Built-In Transform Streams   141
Summary   142

7. Child Processes. 143
Introducing Child Processes   143
The spawn Function   144
Shell Syntax and the exec Function   147
The execFile Function   150
The fork Function   150
Summary   153

8. Testing Node. 155
Assertions and Runners   155
Types of Tests   160
    Unit Tests   161
    Functional Tests   161
    Integration Tests   162
    End-to-End Tests   162
Test Doubles   163
Organizing and Filtering Tests   169
Test-Driven Development   170
Continuous Integration   171
Summary   172

9. Scaling Node. 175
Strategies of Scalability   175
The Cluster Module   176
Primary and Worker Processes   178
Broadcasting Messages   182
Increasing Availability   185
Zero-Downtime Restarts   187

    Handling State   191

Handling State 191

Process Managers 192

Summary 193

10. **Practical Node**.. **195**

Code Quality Tools 195

  Prettier 197

  ESLint 198

  Other Tools 199

Module Bundlers 200

Task Runners 203

Frameworks 205

JavaScript Transpilers 209

TypeScript 210

Summary 216

Index... **217**

Preface

I've been using Node.js since its early days, and it has never failed me. With every piece of code I've written for Node, my appreciation for it has only increased. With every new skill I've developed for Node, I've felt the productivity gain.

Node.js is nothing short of revolutionary. It's a great platform with impressive power. Once you get comfortable with its fundamentals and how it handles asynchrony, the rest is easy. You'll get better with it quickly, and you'll be able to build and scale back-end services faster than you'd imagine.

Who Should Read This Book

This book is my attempt at helping you learn Node.js efficiently. It naturally dips into a few JavaScript concepts, but in general, you need a good basic understanding of the JavaScript language to get the most value out of this book.

If you're not comfortable working with JavaScript objects, functions, operators, and iterators, reading an introductory book about JavaScript before this book would help.

This is the book that I wished existed when I started learning Node.js. At that time, I was mainly focusing on the frontend. Naturally, this book is a good fit for a frontend developer wanting to expand their experience to the backend.

Why I Wrote This Book

When it comes to learning Node.js, many tutorials, books, and courses tend to focus on the libraries and tools available within the Node.js ecosystem, rather than the Node.js runtime environment itself. They prioritize teaching how to utilize popular Node.js libraries and frameworks, instead of starting from the native capabilities of Node.js.

This approach is understandable because Node.js is a low-level runtime environment. It does not offer comprehensive solutions but rather a collection of small essential modules that makes creating solutions easier and faster. For example, a full-fledged web server will have options like serving static files (like images, CSS files, etc.). With the Node.js built-in `http` module, you can build a web server that serves binary data, and with the Node.js built-in `fs` module, you can read the content of a file from the filesystem. You can combine both of these features to serve static assets by using your own JavaScript code. There's no built-in Node.js way to serve static assets under a web server.

Popular Node.js libraries that are not part of Node.js itself (such as Express.js, Next.js, and many others with *.js* in their names) aim to provide nearly complete solutions within specific domains. For example, Express.js specializes in creating and running a web server (and serving static assets, and many other neat features). Practically, most developers will not be using Node.js on its own, so it makes sense for educational materials to focus on the libraries offering comprehensive solutions, so learners can skip to the good parts. The common thinking here is that only developers whose job is to write these libraries need to understand the underlying base layer of Node.js.

However, I would argue that a solid understanding of the built-in power of Node.js is essential before utilizing any of its external libraries and tools. Having a deep understanding of Node.js allows developers to make informed decisions when choosing which libraries to use and how to use them effectively. This book is my attempt to prioritize first learning the native capabilities of Node.js and then using that knowledge to efficiently utilize the powerful libraries and tools in its ecosystem.

Conventions Used in This Book

The following typographical conventions are used in this book:

Italic
>   Indicates new terms, URLs, email addresses, filenames, and file extensions.

`Constant width`
>   Used for program listings, as well as within paragraphs to refer to program elements such as variable or function names, databases, data types, environment variables, statements, and keywords.

`Constant width italic`
>   Shows text that should be replaced with user-supplied values or by values determined by context.

This element signifies a tip or suggestion.

This element signifies a general note.

This element indicates a warning or caution.

Using Code Examples

Supplemental material (code examples, exercises, etc.) is available for download at *https://oreil.ly/EfficientNodeCode*.

If you have a technical question or a problem using the code examples, please send email to *support@oreilly.com*.

This book is here to help you get your job done. In general, if example code is offered with this book, you may use it in your programs and documentation. You do not need to contact us for permission unless you're reproducing a significant portion of the code. For example, writing a program that uses several chunks of code from this book does not require permission. Selling or distributing examples from O'Reilly books does require permission. Answering a question by citing this book and quoting example code does not require permission. Incorporating a significant amount of example code from this book into your product's documentation does require permission.

We appreciate, but generally do not require, attribution. An attribution usually includes the title, author, publisher, and ISBN. For example: "*Efficient Node.js* by Samer Buna (O'Reilly). Copyright 2025 Samer Buna, 978-1-098-14519-4."

If you feel your use of code examples falls outside fair use or the permission given above, feel free to contact us at *permissions@oreilly.com*.

O'Reilly Online Learning

O'REILLY® For more than 40 years, *O'Reilly Media* has provided technology and business training, knowledge, and insight to help companies succeed.

Our unique network of experts and innovators share their knowledge and expertise through books, articles, and our online learning platform. O'Reilly's online learning platform gives you on-demand access to live training courses, in-depth learning paths, interactive coding environments, and a vast collection of text and video from O'Reilly and 200+ other publishers. For more information, visit *https://oreilly.com*.

How to Contact Us

Please address comments and questions concerning this book to the publisher:

O'Reilly Media, Inc.
1005 Gravenstein Highway North
Sebastopol, CA 95472
800-889-8969 (in the United States or Canada)
707-827-7019 (international or local)
707-829-0104 (fax)
support@oreilly.com
https://oreilly.com/about/contact.html

We have a web page for this book, where we list errata, examples, and any additional information. You can access this page at *https://oreil.ly/EfficientNodeJS*.

For news and information about our books and courses, visit *https://oreilly.com*.

Find us on LinkedIn: *https://linkedin.com/company/oreilly-media*.

Watch us on YouTube: *https://youtube.com/oreillymedia*.

Acknowledgments

I am deeply grateful to the many incredible people whose efforts helped shape and refine this book. A heartfelt thank you to the O'Reilly Media team for their patience, guidance, and unwavering support throughout the writing process. I am especially thankful to Jeff Bleiel, my development editor, whose thoughtful insights greatly improved the clarity and presentation of this work. To Amanda Quinn, my editor; Miah Sandvik, my copyeditor; and Gregory Hyman, my production editor: your contributions are greatly appreciated.

I also want to thank the talented software developers who reviewed drafts of this book and provided invaluable feedback. Your suggestions were instrumental in improving the content. Special thanks to Hazem Twair for spotting issues I would have missed, and to Tamas Piros and Aniket Wattamwar for their meticulous technical proofreading. Your input made this book stronger and more polished.

To my mentors and peers in the Node.js community: I am forever grateful for your wisdom, inspiration, and encouragement.

To everyone who contributed: your dedication and expertise have elevated this book far beyond what I could have achieved on my own. Thank you for your help—it means the world to me!

CHAPTER 1

Node Fundamentals

Node is an open source, cross-platform runtime environment in which developers can create backend services using the JavaScript language. It's built on top of V8, the JavaScript engine of the Chrome web browser, and it has dozens of built-in modules that are designed to be used asynchronously with an event-driven approach that's commonly known as the non-blocking model. Node developers can use events and handler functions to efficiently perform multiple operations in parallel, without having to deal with the complexity of multiple processes and threads.

There's a lot to unpack here, and that's what we will be doing in this first chapter. We'll start with an introduction to Node, how it works, and why it's popular. We'll learn the basics of the Node CLI, how to use modules and packages, and how to perform synchronous and asynchronous operations. We'll discuss the fundamentals of Node's event-driven, non-blocking model and learn how callbacks, promises, and events can be used to handle the result of an asynchronous operation.

> Throughout the book, I use the term *Node* instead of *Node.js* for brevity. The official name of the runtime environment is Node.js, but referring to it as just Node is common.

Introducing Node

Ryan Dahl started the Node project in 2009 after he was inspired by the performance of the V8 JavaScript engine in the Google Chrome web browser. V8 uses an *event-driven model*, which makes it efficient at handling concurrent connections and requests. Ryan wanted to bring this same high-performance, event-driven architecture to server-side applications. The event-driven model is the first and most important concept you need to understand about Node (and the V8 engine as well). I'll explain it briefly in this chapter, and we'll expand on it in Chapter 3.

> I decided to give Node a spin and learn more about it after watching the presentation Ryan Dahl gave to introduce it. I think you'll benefit by starting there as well. Search YouTube for "Ryan Dahl introduction to Node" (*https://oreil.ly/TACAt*). Node has changed significantly since then, so don't focus on the examples but rather the concepts and explanations.

In its core, Node enables developers to use the JavaScript language on any machine without needing a web browser. Node is usually defined as "JavaScript on backend servers." Before Node, that was not a common or easy thing. JavaScript was mainly a frontend thing.

However, this definition isn't completely accurate. Node offers a lot more than the ability to execute JavaScript on servers. In fact, the actual execution of JavaScript is done by the V8 JavaScript engine, not Node. Node is just an interface to V8 when it comes to executing JavaScript code.

V8 is Google's open source JavaScript engine that can compile and execute JavaScript code. It's used in Node as well as in Chrome and a few other browsers. It's also used in Deno, the new JavaScript runtime that was created by Ryan Dahl in 2018.

> There are other JavaScript engines, like SpiderMonkey, which is used by Firefox, and JavaScriptCore, which is used by the Safari web browser and in Bun, an all-in-one JavaScript runtime, package manager, and bundler.

Node is better defined as a server runtime environment that wraps V8 and provides modules to help developers build and run efficient software applications with JavaScript.

The key word in this definition is *efficient*. Node adopts and expands on the same event-driven model that V8 has. Most of Node's built-in modules are event-driven and can be used asynchronously without blocking the main *thread* of execution that your code runs in.

A thread is basically a small process within a larger one. A process can create multiple threads of execution that are each associated with a CPU core. Threads can share memory and resources within the larger process.

In multithreaded programming, slow operations are executed in separate threads. In Node, you get a single main thread for your code, and all the slow operations are executed outside of that main thread, asynchronously.

You need to read the content of an external file? You can do that asynchronously without blocking the single main thread. You need to start a web server? Work with network sockets? Parse, compress, or encrypt data? Every low-level slow operation has an asynchronous API for you to use without blocking your other operations.

You don't need to deal with multiple threads to do things in parallel in Node. You don't waste resources on manual threads being idle waiting on slow operations. You code in one thread and use asynchronous APIs, and Node takes care of executing the asynchronous operations efficiently outside of your main thread.

Any code that needs to be executed after a slow operation can be managed with *events* and *event handlers*. An event is a signal that something has happened and a certain action needs to be performed. The action can be defined in an event handler function that gets associated with the event. Every time the event is signaled, its handler function will be executed. That's basically the gist of what *event-driven* means.

We'll expand on these important concepts once we learn the basics of running Node code and using its modules and packages.

The JavaScript Language

After considering programming languages like Python, Lua, and Haskell, Ryan Dahl picked the JavaScript language for Node because it was a good fit. It's simple, flexible, and popular, but more importantly, JavaScript functions are first-class citizens that we can treat like any other objects (numbers or strings). We can store them in variables, pass them to other functions via arguments, and even return them from other functions, all while preserving their state. Node leveraged that to implement its handling of asynchronous operations.

> Despite JavaScript's historical problems, I believe it's a decent language today that can be made even better by using TypeScript (which we will discuss in Chapter 10).

Besides simplifying the implementation of asynchronous operations, the fact that JavaScript is the programming language of browsers gave Node the advantage of having a single language across the full stack. There are some subtle but great benefits to that:

- One language means less syntax to keep in your head, fewer APIs and tools to work with, and fewer mistakes overall.

- One language means better integrations between your frontend code and your backend code. You can actually share code between these two sides. For example, you can build a frontend application with a JavaScript framework like React, then use Node to render the same components of that frontend application on the server and generate initial HTML views for the frontend application. This is known as server-side rendering (SSR), and it's something that many Node frontend frameworks offer out of the box.

- One language means teams can share responsibilities among different projects. Projects don't need a dedicated team for the frontend and a different team for the backend. You would also eliminate some dependencies between teams. A full stack project can be assigned to a single team, *The JavaScript People*; they can develop APIs, they can develop web and network servers, they can develop interactive websites, and they can even develop mobile and desktop applications. Hiring JavaScript developers who can contribute to both frontend and backend applications is attractive to employers.

Executing Node Code

If you have Node installed on your computer, you should have the commands node and npm available in a terminal. If you have these commands, make sure the Node version is a recent one (20.x or higher). You can verify that by opening a terminal and running the command node -v.

If you don't have the node command, you'll need to download and install Node from the Node website (*https://nodejs.org*). The installation process is straightforward and should only take a few minutes.

For macOS users, Node can also be installed using the Homebrew package manager with the command:

```
$ brew install node
```

> Throughout this book, I use the $ sign to indicate a command line to be executed in a terminal. The $ sign is not part of the command. It's a common prompt character in terminals.

Another option to install Node is using Node Version Manager (NVM). NVM allows you to run multiple versions of Node and switch between them easily. You might need to run a certain project with an older version of Node, and use the latest Node version with another project. NVM works on Mac and Linux, and there's a Windows option as well, called *nvm-windows*.

Node on Windows

All the examples I will be using in this book are designed for a macOS environment and should also work for a Linux-based OS. On Windows, you need to switch the commands I use with their Windows alternatives.

I don't recommend using Node on Windows natively unless it's your only option. If you have a modern Windows machine, one option that might work a lot better for you is to install the Windows subsystem for Linux. This option will give you the best of both worlds. You'll have your Windows OS running Linux without needing to reboot. You can even edit your code in a Windows editor and execute it in Linux!

If you're using NVM, install the latest version of Node with the command:

```
$ nvm install node
```

> Major Node versions are released frequently. When a new version is released, it enters a *Current* release status for six months to give library authors time to make their libraries compatible with the new version. After six months, odd-numbered releases (19, 21, etc.) become unsupported, and even-numbered releases (18, 20, etc.) move to *Active LTS* (long-term support) status. LTS releases typically guarantee that critical bugs will be fixed for a total of 30 months. Production applications should use only Active LTS releases.

Once you have the node command ready, open a terminal and issue the command on its own without any arguments. This will start a Node *REPL* session. REPL stands for Read-Eval-Print-Loop. It's a convenient way to quickly test simple JavaScript and Node code. You can type any JavaScript code in a REPL session. For example, try a Math.random() line, as shown in Figure 1-1.

```
TERMINAL                                         node  + ∨  ⊡  🗑  ⋯

○  $ node
   >
   > Math.random()
   0.08949350637305153
   >
   > |
```

Figure 1-1. Node's REPL mode

Node will read the line, evaluate it, print the result, and loop over these three things for everything you type until you exit the session (which you can do with Ctrl + D). Note how the *Print* step happened automatically. We didn't need to add any instructions to print the result. Node will just print the result of each line you type. This is not the case when you execute code in a Node script. Let's do that next.

We'll learn more about Node's REPL mode in Chapter 2.

Create a new directory for the exercises of this book, and then cd into it:

```
$ mkdir efficient-node
$ cd efficient-node
```

Open up your code editor and create a file named *test.js*. Put the same Math.random() line into it:

```
Math.random();
```

Now to execute that script, in the terminal type the following command:

```
$ node test.js
```

You'll notice that the command will basically do nothing. That's because we did not *output* anything in that script. To output something, you can use the global console object, which is similar to the one available in browsers. Here is an example:

```
console.log(
  Math.random()
);
```

Executing *test.js* now will output a random number, as shown in Figure 1-2. In this simple example, we're using both JavaScript (Math object) and an object from the Node API (console). The console.log method writes the value of its arguments to the default standard output stream (*stdout*) of the running process.

> The console object is one of many top-level *global scope* objects that we can access in Node without needing to declare any dependencies. Similar to how the global window object in browsers can be accessed with the globalThis property, in Node, the globalThis property is the global object and the console object is part of it. All properties of globalThis can be accessed directly: for example, console.log instead of globalThis.console.log (which also works).

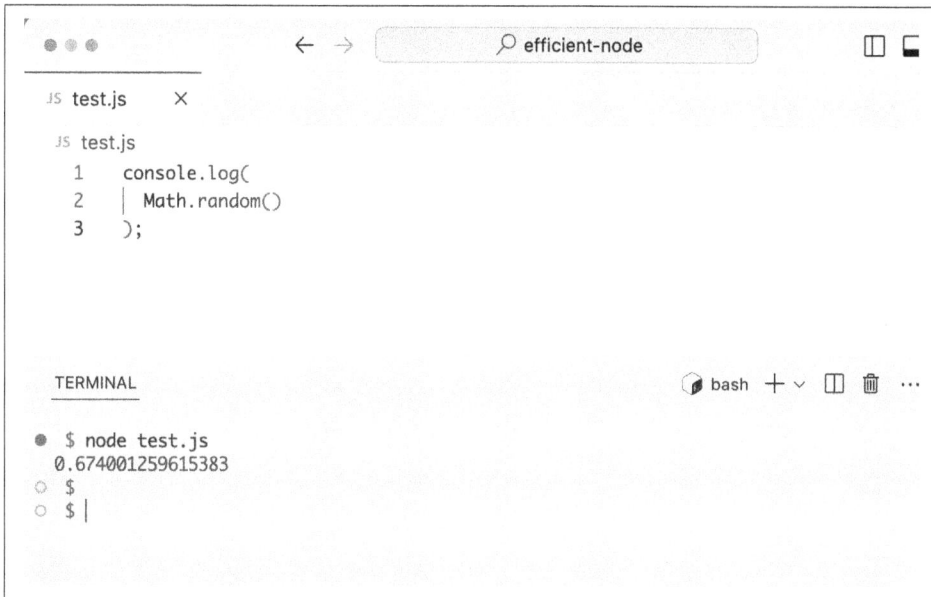

Figure 1-2. Executing a Node script

Using Built-In Modules

You can create a simple web server in Node using its built-in `node:http` module.

Create a *server.js* file and write the following code in it:

```
// Basic Web Server Example

const { createServer } = require('node:http');

const server = createServer((req, res) => {
  res.writeHead(200, { 'Content-Type': 'text/plain' });
  res.end('Hello World');
});

server.listen(3000, '127.0.0.1', () => {
  console.log('Server is running...');
});
```

This is Node's version of a "Hello World" example. You don't need to install anything to run this script. This is all Node's built-in power.

When you execute this script, Node creates a web server and runs it on *http://127.0.0.1:3000*, as shown in Figure 1-3.

```
JS server.js    X

JS server.js > ...

  3    const { createServer } = require('node:http');
  4
  5    const server = createServer((req, res) => {
  6      res.writeHead(200, { 'Content-Type': 'text/plain' });
  7      res.end('Hello World');
  8    });
  9
 10    server.listen(3000, '127.0.0.1', () => {
 11      console.log('Server is running...');
 12    });
 13
```

TERMINAL

$ node server.js
Server is running...

127.0.0.1:3000   ×   +

← → C ⓘ 127.0.0.1:3000   ☆ ⎙ ⬤

Hello World

Figure 1-3. A basic web server

> Note how in this example, the Node process continues to run indefinitely (unless it encounters any unexpected errors). This is because it has work to do in the background. It needs to wait for users to request data and send them a response when they do.

While this web server example is a basic one, it has a few important concepts to understand. Let's go over it in detail.

The require function is part of Node's original dependency management method. It allows a module (like *server.js*) to use the features of another module (like node:http). By requiring the node:http module, the *server.js* module now *depends* on it. It cannot run without it.

Another way for one module to use the features of another module is with an ES modules import statement. ES modules are the modern ECMAScript standard for working with modules in JavaScript. We'll be mostly using them in this book as they are the preferred module system to use in Node today. However, it's good to learn the original module management system (which is known as CommonJS) as many Node projects and libraries are built using that legacy system, and it's very likely that you will have to deal with them even if you're starting a project from scratch.

There are many libraries that you can use to create a web server, but `node:http` is part of Node itself. You don't need to install anything to use it, but you do need to require (or import) it.

> In a Node's REPL session, built-in modules (like `node:http`) are available globally without needing to require them. This is not the case with executable scripts. You can't use modules (including built-in ones) without declaring them first.

You don't need to load everything in a module when you require it. You can pick and choose. This example loads only the `createServer` function, which is one of many functions and other objects that are available in `node:http`.

We invoke `createServer` to create a server object. Its argument is another function that is known as the `RequestListener`. Don't worry about the syntax in this example; focus on the concepts.

A listener function in Node is associated with a certain event, and it gets executed when its event is triggered. In this example, Node will execute the `RequestListener` function every time there is an incoming connection request to the web server. That's the event associated with this listener function.

The listener function receives two arguments:

The request object (named `req` *in this example)*
    You can use this object to read data about incoming requests. For example, what URL is being requested, or what is the IP address of the client that's making a request.

The response object (named `res` *in this example)*
    You can use this object to write things back to the requester. It's exactly what this simple web server is doing. It's setting the response status code to `200` to indicate a successful response, and the `Content-Type` header to `text/plain`. Then it's writing back the `Hello World` text using the end method on the `res` object.

The `createServer` function only creates the server object. It does not activate it. To activate this web server, you need to invoke the `listen` method on the created server.

The `listen` method accepts many arguments, like what OS port and host to use for this server. The last argument for it is a function that will be invoked once the server is successfully running on the specified port. This example prints a console message to indicate that the server is running successfully at that point.

Both functions received by the `createServer` and `listen` methods are examples of handler functions that are associated with events related to an asynchronous operation. We'll learn how these events and their handler functions are managed in Chapter 3.

> Note how I use a `node:` prefix when working with the built-in modules in Node. This is a helpful practice to distinguish them from external modules and identify their built-in nature. Since it's required for a few modules (like `node:test`), it's good to be consistent and use it for all modules.

To stop the web server, press Ctrl + C in the terminal where it's running.

Using Packages

Node's package manager is npm. It's a simple CLI that lets us install and manage external *packages* in a Node project. An npm package can be a single module or a collection of modules grouped together and exposed with an API. We'll talk more about npm and its commands and packages in Chapter 5. Here, let's just look at a simple example of how to install and use an npm package.

Let's use the popular `lodash` package, which is a JavaScript utility library with many useful methods you can run on numbers, strings, arrays, objects, and more.

First, you need to download the package. You can do that using the `npm install` command:

```
$ npm install lodash
```

This command will download the `lodash` package from the npm registry, and then place it under a *node_modules* folder (which it will create if it's not there already). You can verify with an `ls` command:

```
$ ls node_modules
```

You should have a folder named *lodash* in there.

Now in your Node code, you can `require` the `lodash` module to use it. For example, `lodash` has a `random` method that can generate a random number between any two numbers you pass to it. Here's an example of how to use it:

```
const _ = require('lodash');

console.log(
  _.random(1, 99)
);
```

When you run this script, you'll get a random number between 1 and 99, as shown in Figure 1-4.

```
JS index.js    ×

JS index.js > ...
  1    const _ = require('lodash');
  2
  3    console.log(
  4    |  _.random(1, 99)
  5    );
  6



TERMINAL                                            bash + ∨  ⊓  🗑  ⋯

●  $ node index.js
   56
○  $
○  $ |
```

Figure 1-4. Using an npm package

> The underscore (_) is a common name to use for lodash, but you can use any name.

Since we called the require method with a non-built-in module, lodash, Node will look for it under the *node_modules* folder, and thanks to npm, it'll find it there.

In a team Node project, when you make the project depend on an external package like this, you need to let other developers know of that dependency. You can do so in Node using a *package.json* file at the root of the project.

When you npm install a module, the npm command will also list the module and its current version in *package.json*, under a dependencies section. Look at the *package.json* file that was autocreated when you installed the lodash package and see how the lodash dependency was documented (see Figure 1-5).

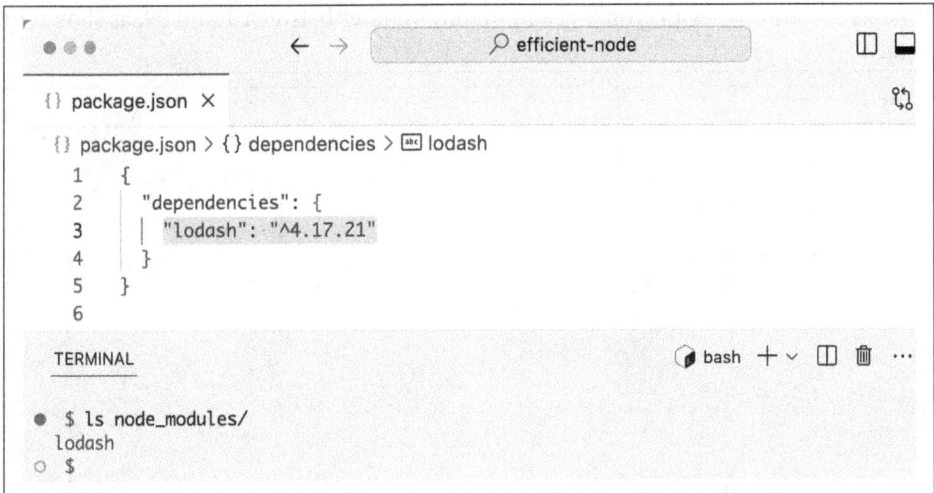

Figure 1-5. The package.json *file*

The *package.json* file can also contain information about the project, including the project's name, version, description, and more. It can also be used to specify scripts that can be run from the command line to perform various tasks, like building or testing the project.

Here's an example of a typical *package.json* file:

```
{
  "name": "efficient-node",
  "version": "1.0.0",
  "description": "A guide to learning Node.js",
  "license": "MIT",
  "scripts": {
    "start": "node index.js"
  },
  "dependencies": {
    "lodash": "^4.17.21"
  }
}
```

You can interactively create a *package.json* file for a new Node project using the npm init command:

```
$ npm init
```

This command will ask a few questions, and you can interactively supply your answers (or press Enter to keep the defaults, which often are good because the command tries to detect what it can about the project).

Try to `npm install` new packages (for example, `chalk`) and see how it gets listed as a dependency in your *package.json* file. Then `npm uninstall` the package and see how it gets removed from *package.json*.

Your *package.json* file will eventually list many dependencies. When other developers pull your code, they can run the command `npm install` without any arguments, and npm will read all the dependencies from *package.json* and install them under the *node_modules* folder.

Some packages are needed only in a development environment, not in a production environment. The ESLint package is an example of that. You can instruct the `npm install` command to list a package as a development-only dependency by adding the `--save-dev` argument (or `-D` for short):

```
$ npm install -D eslint
```

This will install the `eslint` package in the *node_modules* folder, and document it as a development dependency under a `devDependencies` section in *package.json*. This is where you should place things like your testing framework, your formatting tools, or anything else that you use only while developing your project.

> In addition to `dependencies` and `devDependencies`, a *package.json* file can also specify `optionalDependencies` for packages that are optional, and `peerDependencies` for packages that need to work alongside other packages but do not directly depend on them. Peer dependencies are needed only by package authors.

If you take a look at what's under *node_modules* after you install `eslint`, you'll notice that there are dozens of new packages there (see Figure 1-6).

The `eslint` package depends on all these other packages. Be aware of these indirect dependencies in the future. By depending on one package, a project is indirectly depending on all of that package's dependencies, and the dependencies of all the subdependencies, and so on. With every package you install, you add a tree of dependencies.

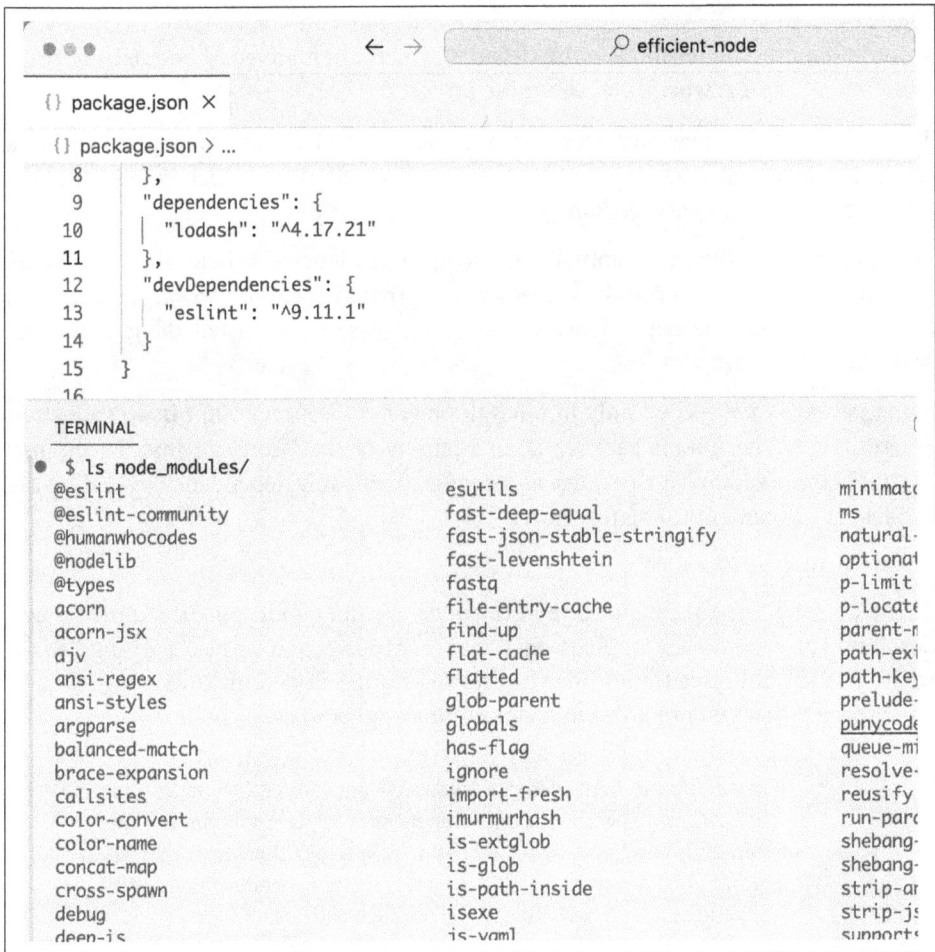

Figure 1-6. npm packages and their indirect dependencies

Some packages can also be installed (and configured) directly with the `init` command. ESLint is an example of a package that needs a configuration file before you can use it. The following command will install ESLint and create a configuration file for it (after asking you a few questions about your project):

```
$ npm init @eslint/config@latest
```

In a production machine, development dependencies are usually ignored. The npm install command has a --production flag to make it ignore them. You can also use the NODE_ENV environment variable and set it to production before you run the npm install command. We'll learn more about Node environments and variables in Chapter 2.

ES Modules

Node has two different module loaders. The default one is the CommonJS module loader, and we saw an example of it that uses the require function.

The other one is the JavaScript-native ES module loader that is supported in modern browsers as well. In ES modules, we use import statements to declare a module dependency and export statements to make one module's features available for other modules to use.

One important difference between these two module systems is that CommonJS modules get loaded dynamically at runtime, while ES module dependencies are determined at compile time, allowing them to be statically analyzed and optimized. For example, with ES modules we can easily find what code is not being used and exclude it from the compiled version of the application.

While the two module types can be used together, you need to be careful about mixing them. CommonJS modules are synchronous while ES modules are asynchronous.

To see ES modules in action, let's expand on the basic web server example code and split it into two files: one to create the web server, and one to run it.

The simplest way to use ES modules in Node is to save files with a *.mjs* file extension instead of a *.js* extension. This is because by default, Node assumes that all *.js* extensions are using the CommonJS module system. This is configurable though.

To make Node treat all *.js* files as ES modules, you can add a type key in *package.json* and give it the value of module (the default value for it is commonjs). You can do that manually or with this command:

```
$ npm pkg set type=module
```

With that, you can now use ES modules with the *.js* extension.

> Regardless of what default module type you use, Node will always assume a *.mjs* file is an ES module file and a *.cjs* file is a CommonJS module file. You can import a *.cjs* file into an ES module, and you can import a *.mjs* file into a CommonJS module.

Let's modify the basic web server example to use ES modules. In the *server.js* file, write the following code:

```
import { createServer } from 'node:http';

export const server = createServer((req, res) => {
  res.writeHead(200, { 'Content-Type': 'text/plain' });
  res.end('Hello World');
});
```

Note the use of import/export statements. This is the syntax for ES modules. You use import to declare a module dependency and export to define what other modules can use when they depend on your module.

In this example, the *server.js* module exports the server object, enabling other modules to import it and depend on it.

To use the *server.js* exported objects in other modules, we use another import statement. In an *index.js* file, write the following code:

```
import { server } from './server.js';

server.listen(3000, () => {
  console.log('Server is running...');
});
```

The ./ part in the import line signals to Node that this import is a relative one. Node expects to find the *server.js* file in the same folder as *index.js*. You can also use a ../ to make Node look for the module up one level, or ../../ for two levels, and so on. Without ./ or ../, Node assumes that the module you're trying to import is either a built-in module or a module that exists under the *node_modules* folder.

With this code, the *index.js* module depends on the *server.js* module and uses its exported server object to run the server on port 3000.

Execute *index.js* to start the web server and test it, as illustrated in Figure 1-7.

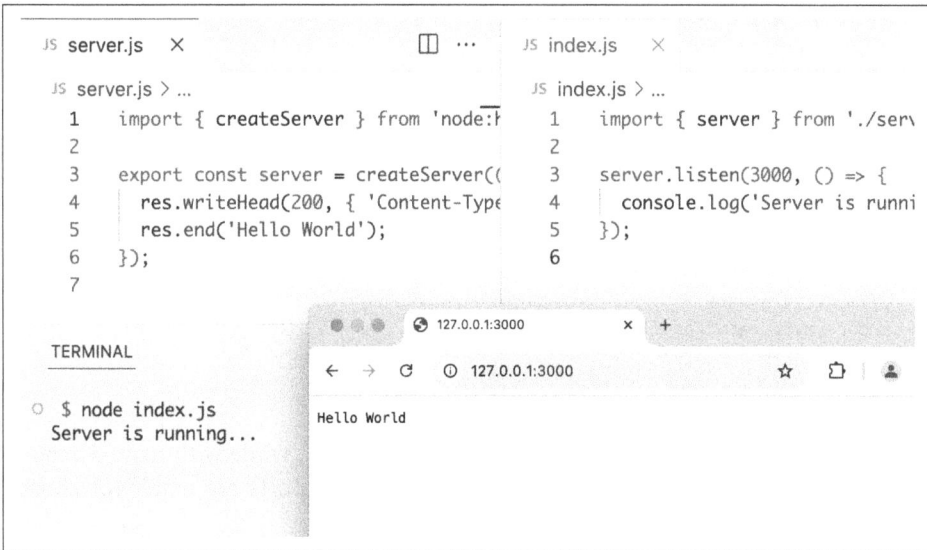

Figure 1-7. An ES modules web server

The `export object` syntax is known as *named exports* and it's great when you need to export multiple things in a module. You can use the `export` keyword to prefix any object, including functions, classes, and destructured variables:

```
export function functionName() { ... }
export class ClassName { ... }
export const [name1, name2]
```

You can also use one export keyword, usually at the end of a module, to export all named objects together:

```
export {
  functionName,
  ClassName,
  name1,
  name2,
  // ...
};
```

You can import these named exports individually, or use the `*` `as` syntax to import all of them:

```
// To import named exports individually:
import { functionName, name1 } from './module'

// To import all named exports:
import * as serviceName from './module'
```

```
// Then access named exports as:
// serviceName.functionName
// serviceName.name1
```

In addition to the named export syntax, ES modules also have a *default export* syntax:

```
// To export the server object
// as the default export in server.js:
export default server;

// To import it, you need to
// specify a name:
import myServer from './server.js';

// Or:
import { default as myServer} from './server.js';
```

Note that to import a default export, you have to name it, while with named exports you don't (although you can if you need to). Named exports are better for consistency, discoverability, and maintainability.

> Since a module might not have a default export, you can always use the * as syntax since it works with or without a default export.

The export/import keywords support other syntax like renaming and exporting from other modules, but in general my recommendation is to avoid using default exports; always use named exports, and keep them simple and consistent. For example, I always specify my named exports as the last line of the module with a single export { ... } statement.

An Analogy for Node and npm

Real-life analogies can sometimes help us understand coding concepts.

One of my favorite analogies about coding in general is how it can be compared to writing cooking recipes. The recipe in this analogy is the program, and the cook is the computer.

In some recipes, you can use premade items like a cake mix or a special sauce. You would also need to use tools like a pan or a strainer. When compared to coding, you can think of these premade items and tools as the packages of code written by others, which you can just download and use.

Extending on this analogy, you can think of an npm registry as the store where you get your premade items and tools for your coding recipes.

But what exactly is Node's place in this analogy?

I like to think of it as the kitchen! It allows you to execute lines in your coding recipes by using built-in tools, like the stove and the sink.

Now imagine trying to cook without these built-in tools in your kitchen. That would make the task a lot harder, wouldn't it?

Asynchronous Operations

While static import declarations are preferable for loading initial dependencies, there are many cases where you will need to import modules dynamically. For example:

- When a module is slowing the loading of your code and it's not needed until a later time
- When a module does not exist at load time
- When you need to dynamically construct the name of the module to import
- When you need to import a module conditionally

For these cases, we can use the `import()` function. It's similar to the `require()` function, but it's asynchronous.

Let's think about an example where we need to read the content of a file before starting the basic web server. We can simulate the file reading delay using a simple `setTimeout` function.

Timer functions can be used to delay the execution of a function or make it repeat regularly. If you're not familiar with them, the next sidebar explains them with examples.

Since we don't need the *server.js* module until a later point in time, we can import it with the `import()` function when we are ready for it:

```
setTimeout(async () => {
  const { server } = await import('./server.js');

  server.listen(3000, () => {
    console.log('Server is running...');
  });
}, 5_000);
```

If you execute this code, the Node process will wait five seconds. It will then dynamically import the *server.js* module and use it to start the server.

This example introduces the important `Promise` object concept and how to consume it with the `async/await` syntax. This is the modern JavaScript syntax to handle asynchronous operations. We'll learn more about promises and the `async/await` syntax in the next section.

> Dynamic import expressions can be used in CommonJS modules to load ES modules.

Timer Functions

Timer functions in Node like `setTimeout` and `setInterval` behave similarly to how they do in browser environments. A timer function receives a function as an argument:

```
const printGreeting = () => console.log('Hello');

setTimeout(printGreeting, 4_000);
```

The `printGreeting` function will be invoked once after four seconds. To invoke it multiple times, you can use the `setInterval` function. Replacing `setTimeout` with `setInterval` in this example will make Node print the "Hello" message every four seconds, forever.

Timer functions can be canceled once they are defined. When you call a timer function, you get back a unique timer ID. You can use that timer ID to cancel the scheduled timer. We can use `clearTimeout(timerId)` to stop timers started by `setTimeout`, and `clearInterval(timerId)` to stop timers started by `setInterval`:

```
const timerId = setTimeout(
  () => console.log('Hello'),
  0,
);

clearTimeout(timerId);
```

In this example, even though we started a timer to print a message after zero milliseconds, the message will *not* be printed at all because we canceled the asynchronous timer operation after it was defined.

The Non-Blocking Model

When you need to perform a slow operation in your code (like reading a file from the filesystem), you'll need a way to handle the output of that slow operation.

Let's simulate a slow operation function with a long-running `for` loop:

```
function slowOperation() {
  for (let i = 0; i <= 1e9; i++) {
    // ...
  }
  return { success: true };
}
```

The `slowOperation` function might return data successfully or throw an error (like a timeout error). Here's a simple example function to handle its output:

```
function handlerFunction(output) {
  if (!output.success) {
    // Something went wrong
  }
  // Do something with output
}
```

Here's how we would use the `slowOperation` function to pass its output to its `handlerFunction`:

```
const output = slowOperation();
handlerFunction(output)

console.log('Hello');
// Other operations
```

The problem with this is that the slow operation will block the execution of all other operations that follow it. The `console.log` operation will wait until both `slow Operation` and `handlerFunction` are done executing.

Since JavaScript functions can be passed as arguments, we can design `slowOperation` to invoke its handler function once it's done, using the following pattern:

```
function slowOperation(callbackFunction) {
  for (let i = 0; i <= 1e9; i++) {
    // ...
  }
  callbackFunction({ success: true });
}

slowOperation(
  (output) => handlerFunction(output)
);

// Other operations
```

Now, we can make the `slowOperation` run in a different thread so that the other operations in the main thread are not blocked. This is known as the *callback pattern* and it's the original implementation of handling asynchronous operations in Node. A callback function gets called at a later point in time once the slow operation is done.

The `setTimeout` function is the simplest example of an asynchronous function that follows the callback pattern:

```
setTimeout(
  function callback() {
    console.log('World');
  },
  2_000, // delay is in ms
);

console.log('Hello');
```

The `setTimeout` function itself is not part of JavaScript. It's implemented in the run-time environment like Node (or the browser). What gets executed by the JavaScript engine is its callback function.

> You can think of the callback pattern as a simple method for performing an asynchronous operation with a handler function and a built-in event. For `setTimeout`, the built-in event is a time delay. The handler function is what we pass to `setTimeout` as a callback function.

The actual timer operation is handled in a separate thread so that it does not block the main thread. That's why the output of this will be as follows:

```
Hello
World
```

The operation that follows the timer operation was executed first. Then, once the built-in event was triggered (two seconds passed by), the callback function that internally gets associated with this timer event was executed by V8.

This would also be the output when the timer delay is zero. All asynchronous operations, no matter how fast they are, get removed from the main thread immediately, processed internally in Node, and return to the main thread (via their callbacks) once all the other synchronous operations are done.

Zero-milliseconds delayed code is a way to *schedule* code to be invoked when all the synchronous code defined after it is done executing. This is a good simple example of the non-blocking nature of Node. If we define the long-running `for` loop in the callback of a `setTimeout` function with a delay of zero, we are basically scheduling that loop to execute after all the synchronous operations that come after it are done:

```
setTimeout(
  () => {
    for (let i = 0; i <= 1e9; i++) {
      // ...
    }
  },
  0,
);

console.log('Hello');
```

The "Hello" message will print first here, then the long-running loop will be executed.

A general observation about timer functions is that their delays are not guaranteed to be exact, but rather a minimum amount. Delaying a function by 10 milliseconds means that the execution of that function will happen after a minimum of 10 milliseconds, but possibly longer depending on the code that comes after it.

A few years after the success of Node and its use of the callback pattern, `Promise` objects were introduced to the JavaScript language. A `Promise` object represents a value that might be available in the future. This enables us to natively wrap an asynchronous operation as a `Promise` object to which handler functions can be attached and executed later once the promise value is resolved.

Here's the main pattern for `Promise` objects applied to our simple `slowOperation` and `handlerFunction` example:

```
const outputPromise = slowOperation();
outputPromise.then(
  (output) => handlerFunction(output)
);

// Other operations
```

The `node:timers` module has a promise-based `setTimeout` function that can be used with this pattern:

```
import { setTimeout } from 'node:timers/promises';

setTimeout(2_000).then(
  function callback() {
    console.log('World');
  }
);

console.log('Hello');
```

This will be equivalent to the callback-based example. The "Hello" message will be printed immediately, then after a two-second delay, the "World" message will be printed.

The callback and promise patterns can both be used in Node today to use the asynchronous APIs of its built-in modules. Let's look at an example from the node:fs module, which we can use to read the content of a file from the filesystem.

Here's the simplest way to do that:

```
// Reading a file synchronously

import { readFileSync } from 'node:fs';

const data = readFileSync('/Users/samer/.bash_history');

console.log(`Length: ${data.length}`);

console.log(`Process: ${process.pid}`);
```

Figure 1-8 shows the output of running this code.

```
JS readFileSync.js ×

JS readFileSync.js > ...
  3    import { readFileSync } from 'node:fs';
  4
  5    const data = readFileSync('/Users/samer/.bash_history');
  6
  7    console.log(`Length: ${data.length}`);
  8
  9    console.log(`Process: ${process.pid}`);
 10
```

TERMINAL

```
$ node readFileSync.js
Length: 179564
Process: 56097
$
```

Figure 1-8. Reading a file synchronously

Reading a file is an I/O operation, and it's done synchronously in this example. This means it'll block the main thread, and any code that's written after it will have to wait until it's done. Note how the `console.log` statement for the process PID had to wait until after the reading operation was done.

This is bad, especially if you're trying to read a big file. If this code was part of a web server, all incoming requests to that server would have to wait until the main thread is not blocked anymore. We'll test an example of that in Chapter 3.

> An *I/O operation* refers to any communication between a computer program process and its outside world. It typically involves the transfer of data to/from storage devices (like hard drives and memory), peripheral devices (like a keyboard, mouse, or printer), or over a network. I/O operations can be slow, and that's why they are usually run in different processes to not block the main thread of execution.

Performing I/O operations synchronously like this might be OK in a few cases. For example, if you need to read a file one time before you start a web server, or right before you stop it, you can do that synchronously. In most other cases, you want to avoid using any synchronous operations and use only non-blocking ones. Here's how you can read the content of a file asynchronously and avoid blocking the main thread:

```
// Reading a file asynchronously

import { readFile } from 'node:fs';

readFile('/Users/samer/.bash_history', function cb(error, data) {
  console.log(`Length: ${data.length}`);
});

console.log(`Process: ${process.pid}`);
```

Figure 1-9 shows the output of running this code. Note how the `console.log` statement for the process PID was executed before the `console.log` statement for the file data length. The file reading operation did not block the main thread.

This is because the `readFile` method is an asynchronous one. Node does not execute it in the main thread at all. It takes it elsewhere and schedules the execution of its associated callback function right after the reading operation is done.

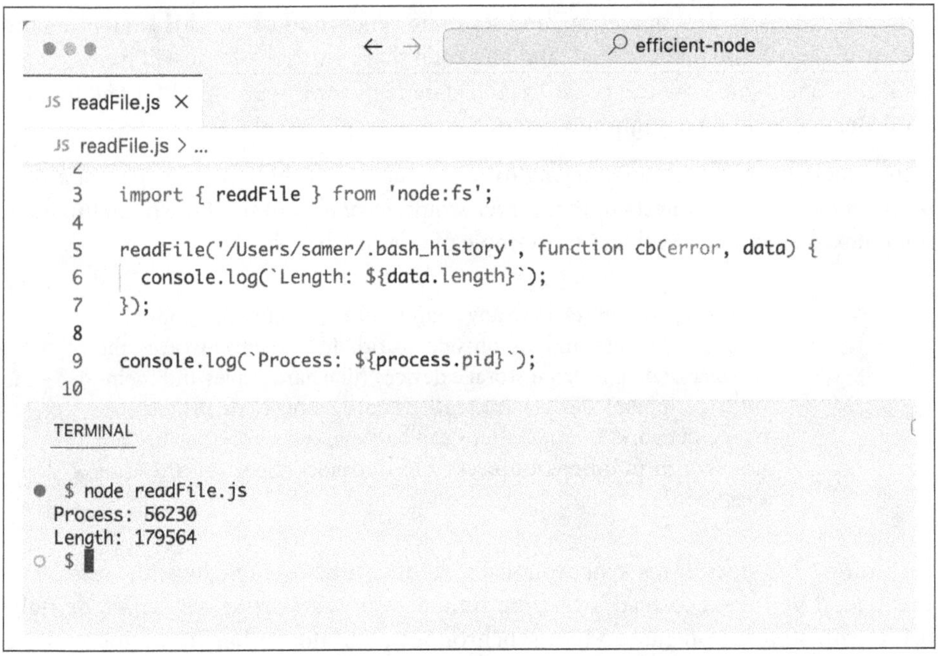

Figure 1-9. Reading a file asynchronously

In this simple example, the callback function is associated with the `readFile` method itself, but internally, that translates to it being associated with an implicit event that gets triggered when the file data is ready.

You'll soon see examples of functions associated with explicit events, either built-in events or user-defined events.

Here's how this example can be converted into using a `Promise` object instead of a callback function:

```
// Reading a file asynchronously with promises

import { readFile } from 'node:fs/promises';

async function logFileLength() {
  const data = await readFile('/Users/samer/.bash_history');
  console.log(`Length: ${data.length}`);
}

logFileLength();

console.log(`Process: ${process.pid}`);
```

Note how the readFile method here is imported from node:fs/promises. This is Node's built-in *promisified* version of the fs module. Executing this promise-based readFile method will return a Promise object.

To access the actual data of this operation, we use the await keyword within an async function. The await keyword pauses the execution of the logFileLength function until the promise is either resolved (success) or rejected (failure). Any function that uses the await keyword becomes an asynchronous function that implicitly returns a Promise object as well.

Figure 1-10 shows the output of running this code example.

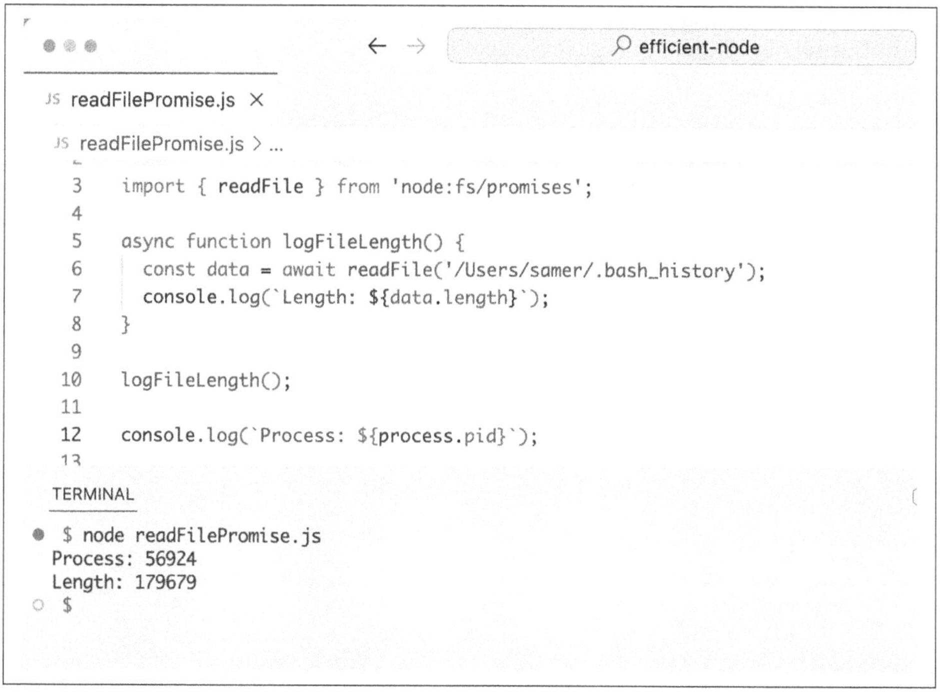

```
JS readFilePromise.js ✕

 JS readFilePromise.js > ...

  3    import { readFile } from 'node:fs/promises';
  4
  5    async function logFileLength() {
  6      const data = await readFile('/Users/samer/.bash_history');
  7      console.log(`Length: ${data.length}`);
  8    }
  9
 10    logFileLength();
 11
 12    console.log(`Process: ${process.pid}`);
 13

TERMINAL

 $ node readFilePromise.js
 Process: 56924
 Length: 179679
 $
```

Figure 1-10. Reading a file with promises

Promise objects make code easier to understand and deal with. Note the similarity of the flow of code when reading a file synchronously and asynchronously with a Promise object. With Promise objects, we get to use Node's non-blocking model without needing to deal with callback functions. Promise objects with the async/await syntax make the code easier to read, especially when there are multiple asynchronous operations that depend on each other. With callbacks, things become a lot more complicated, whereas with promises, we just add more await lines.

We'll learn more about events, callbacks, promises, and the `async`/`await` syntax in Chapter 3.

Node Built-In Modules

Armed with a simple non-blocking model, Ryan Dahl and many early contributors to Node got to work and implemented many low-level modules to offer asynchronous APIs for features like reading and writing files, sending and receiving data over a network, compressing and encrypting data, and dozens of other features.

We've already looked at simple examples of using the `node:http` and `node:fs` modules. To see the list of all built-in modules you get with Node, you can use this line (in a REPL session, as shown in Figure 1-11):

```
require('repl').builtinModules
```

```
TERMINAL                                                       node + v  ▯ ▯ ...

○ $ node
  >
  > require("repl").builtinModules
  [
    'assert',            'assert/strict',        'async_hooks',
    'buffer',            'child_process',        'cluster',
    'console',           'constants',            'crypto',
    'dgram',             'diagnostics_channel',  'dns',
    'dns/promises',      'domain',               'events',
    'fs',                'fs/promises',          'http',
    'http2',             'https',                'inspector',
    'inspector/promises','module',               'net',
    'os',                'path',                 'path/posix',
    'path/win32',        'perf_hooks',           'process',
    'punycode',          'querystring',          'readline',
    'readline/promises', 'repl',                 'stream',
    'stream/consumers',  'stream/promises',      'stream/web',
    'string_decoder',    'sys',                  'timers',
    'timers/promises',   'tls',                  'trace_events',
    'tty',               'url',                  'util',
    'util/types',        'v8',                   'vm',
    'wasi',              'worker_threads',       'zlib'
  ]
  > |
```

Figure 1-11. Node's built-in modules

This is basically the list of things you need to learn to master Node. Well, not all of it. Depending on the version of Node, this list might include deprecated (or soon to be deprecated) modules. You also might not need many of these modules depending on the scope of your work and many other factors. For example, instead of using the native HTTPS capabilities of Node, you can simply put your Node HTTP server behind a reverse proxy like nginx or a service like Cloudflare. Similarly, you would need to learn a module like `wasi` only if you're working with WebAssembly.

Note how a few of these modules are included twice, one with a `/promises` suffix. These are the modules that support both the callback and the promise patterns.

> Not all Node modules will be included in this list. Prefix-only modules and other experimental modules do not show up here. Examples include modules like `node:test`, `node:sea`, and `node:sqlite`. For a full list of modules and their development status, check out the stability overview table in the Node documentation (*https://nodejs.org/docs/latest/api/documentation.html*).

It's good to get familiar with this list now and get a taste of what you can do with Node. Table 1-1 shows some of the important modules with a description of the main tasks you can do with them.

Table 1-1. Key Node modules

Module	Task
`node:assert`	Verify invariants for testing
`node:buffer`	Represent and handle binary data
`node:child_process`	Run shell commands and fork processes
`node:cluster`	Scale a process by distributing its load across multiple workers
`node:console`	Output debugging information
`node:crypto`	Perform cryptographic functions
`node:dns`	Perform name resolutions like IP address lookup
`node:events`	Define custom events and handlers
`node:fs`	Interact with the filesystem
`node:http`	Create HTTP servers and clients
`node:net`	Create network servers and clients
`node:os`	Interact with the operation system
`node:path`	Handle paths for files and directories
`node:perf_hooks`	Measure and analyze applications performance
`node:stream`	Handle large amounts of data efficiently
`node:test`	Create and run JavaScript tests
`node:timers`	Schedule code to be executed at a future time
`node:url`	Parse and resolve URL objects

Module	Task
`node:util`	Access useful utility functions
`node:zlib`	Compress and decompress data

We'll see many examples of using these modules throughout the book. Some of these modules have entire chapters focusing on them.

Node Packages

Node ships with the powerful npm package manager. We did not have a package manager in the JavaScript world before Node, so npm was nothing short of revolutionary. It changed the way we work with JavaScript.

You can build many features in a Node application by using code that's freely available on npm. The npm registry has more than a million packages that you can install and use in your Node servers. The package manager is reliable and comes with a simple CLI. The main `npm` command offers simple ways to install and maintain third-party packages, share your own code, and reuse it too.

> You can install packages for Node from other package registries as well. For example, you can install them directly from GitHub.

Together, npm and Node's module systems make a big difference when you work with any JavaScript system, not just the JavaScript that you execute on backend servers or web browsers. For example, if you have a fancy fridge monitor that happens to run on JavaScript, you can use Node and npm for the tools to package, organize, and manage dependencies, and then bundle your code and ship it to your fridge!

The packages that you can run on Node come in all shapes and forms. Some are small and dedicated to specific programming tasks, some offer tools to assist in the lifecycles of an application, and others help developers every day to build and maintain big and complicated applications. Here are a few examples of some of my favorite tools available from npm:

ESLint

A tool that you can include in any Node applications and use to find problems with your JavaScript code (and, in some cases, automatically fix them). You can use ESLint to enforce best practices and consistent code style, but ESLint can help point out potential runtime bugs too. You don't ship ESLint in your production environments; it's just a tool that can help you increase the quality of your code as you write it.

Prettier

An opinionated code formatting tool. With Prettier, you don't have to manually indent your code, break long code into multiple lines, remember to use a consistent style for the code (for example, always use single or double quotes, always use semicolons, never use semicolons, etc.). Prettier automatically takes care of all of that.

Webpack

A tool that assists with asset bundling. The Webpack Node package makes it very easy to bundle your multifile frontend application into a single file for production, and to compile JavaScript extensions (like JSX for React) during that process. This is an example of a Node tool that you can use on its own. You do not need a Node web server to work with Webpack.

TypeScript

A tool that adds static typing and other features to the JavaScript language. It is useful because it can help developers catch errors before the code is run, making it easier to maintain and scale large codebases. TypeScript's static typing can also improve developer productivity by providing better code autocompletion and documentation in development tools.

All of these tools (and many more) enrich the experience of creating and maintaining JavaScript applications, both on the frontend and the backend. Even if you choose not to host your frontend applications on Node, you can still use Node for its tools. For example, you can host your frontend application with another framework such as Ruby on Rails and use Node to build assets for the Rails asset pipeline.

We will learn more about these tools (and others) in Chapter 10.

Arguments Against Node

Node's approach to handling code in an asynchronous and non-blocking manner is a unique model of thinking and reasoning about code. If you've never done it before, it will feel weird at first. You need time to get your head wrapped around this model and get used to it.

Node's module system was originally built around CommonJS, which has since been largely replaced by the newer ES modules standard in JavaScript. While Node supports both systems, using them together can be confusing, especially for beginners. The differences in how CommonJS and ES modules handle imports and exports can lead to inconsistent code and compatibility issues.

Node developers rely on many third-party libraries and dependencies, and npm stores them all in one large *node_modules* folder, which can become bloated and difficult to manage. It's not uncommon for a Node project to use hundreds of third-party packages, which require management and oversight. As packages are regularly updated or abandoned, it becomes necessary to closely monitor and update all packages used within a project, resolving any version conflicts, replacing deprecated options, and ensuring that your code is not vulnerable to any of the security problems these packages might introduce.

Security in general is one of the strongest arguments against Node. A Node script has unrestricted access to the filesystem, network, and other system resources. This can be dangerous when running third-party code because malicious scripts could exploit these permissions. Node is introducing a new permission model to restrict access to specific resources during execution. You can restrict a Node process from accessing the filesystem, spawning new processes, using worker threads, using native add-ons, and using WebAssembly. However, these restrictions are not enabled by default.

Another limitation in Node is the lack of built-in tools for tasks like validating types, linting, and formatting code. Developers typically have to rely on third-party packages to add these features. While there are plenty of great options, setting up and configuring them can be time-consuming and adds extra steps before you can start coding.

Additionally, Node is optimized for I/O operations and high-level programming tasks, but it may not be the best choice for CPU-bound tasks, such as image and video processing, which require a lot of computational power. Because Node is single-threaded, meaning that it can use only one core of a CPU at a time, performing tasks that require a lot of CPU processing power might lead to performance bottlenecks. JavaScript itself is not the best language for high-performance computation, as it is less performant than languages like C++ or Rust.

Finally, JavaScript, the language you use in Node, has one important argument against it. It is a dynamically typed language, which means objects don't have explicitly declared types at compile time, and they can change during runtime. This is fine for small projects, but for bigger ones, the lack of strong typing can lead to errors that are difficult to detect and debug, and it generally makes the code harder to reason with and maintain.

Summary

Node is a powerful framework for building backend services. It wraps the V8 Java-Script engine to enable developers to execute JavaScript code in a simple way, and it is built on top of a simple, event-driven, non-blocking model that makes it easy for developers to create efficient and scalable applications.

In Node, asynchronous operations are handled with callback functions or `Promise` objects. Callbacks and promises are simple implementations of a one-time event that gets handled with one function. Promises are a better alternative to callbacks as they offer a more readable syntax and can be structured in a way to allow for more control over the code.

The built-in modules in Node provide a low-level framework on which developers can base their applications so that they don't start from scratch. Node's module system allows developers to organize their code into reusable modules that can be imported and used in other parts of the application. Node has a large and active community that has created many popular packages that can be easily integrated into Node projects. These packages can be found and downloaded from the npm registry.

In the next chapter, we'll explore Node's CLI and REPL mode and learn how Node loads and executes modules.

CHAPTER 2

Scripts and Modules

We used the node command briefly in Chapter 1 to explore Node's REPL mode and execute simple scripts. In this chapter, we'll learn how Node loads and executes scripts and modules. We'll start by exploring more of the options, arguments, and environment variables that can be used with the node command, and learning more about what we can do in a REPL session. Then we'll learn about the steps Node takes to load and execute a module.

Node CLI

The node command has many options that can be used to customize its behavior. It also supports arguments and environment variables to further customize what it does, and to pass data from the OS environment to Node's process environment.

Let's take a look. In the terminal, type the following:

```
$ node -h | less
```

This will output the help documentation for the command (one page at a time because we piped the output on the less command). I find it useful to always get familiar with the help pages for the commands I use often:

```
Usage: node [options] [ script.js ] [arguments]
       node inspect [options] [ script.js | host:port ] [arguments]

Options:
  -                           script read from stdin (default if no
                              file name is provided, interactive mode
                              if a tty)
  --                          indicate the end of node options
  --abort-on-uncaught-exception
                              aborting instead of exiting causes a
                              core file to be generated for analysis
```

```
    --build-snapshot          Generate a snapshot blob when the
                              process exits. Currently only supported
                              in the node_mksnapshot binary.
    -c, --check               syntax check script without executing
    --completion-bash         print source-able bash completion
                              script
    -C, --conditions=...      additional user conditions for
                              conditional exports and imports
    --cpu-prof                Start the V8 CPU profiler on start up,
  :
```

The first two lines specify how to use the node command. Anything in square brackets is optional, which means, according to the first line, we can use the node command on its own without any options, scripts, or arguments. That's what we did to start a REPL session. To execute a script, we used the node *script*.js syntax (*script* can be any name there).

What's new here is that there are options and arguments that we can use with the command. Let's talk about these.

> The second usage line is to start a terminal debugging session for Node. While that's sometimes useful, in Chapter 4 I'll show you a much better way to debug code in Node.

In the help page, right after the usage lines, there is a list of all the options that you can use with the command. Most of these options are advanced, but knowing of their existence is a helpful reference. You should scan through this list just to get a quick idea of all the types of things that you can do with the command. Let me highlight a few of the options that I think you should be aware of.

Options and Arguments

The --check option (or -c) lets you check the syntax of a Node script without running that script. An example use of this option is to automate a syntax check before sharing code with others.

The --eval and --print options (or -e and -p) can both be used for executing code directly from the command line. I like the -p option more because it executes and prints (just like in the REPL mode). To use these options, you pass them a string of Node code. Here is an example:

```
$ node -p "Math.random()"
```

This is handy, as you can use it to create your own powerful commands (and alias them if you want). For example, say you need a command to generate a unique random string (to be used as a password maybe). You can leverage Node's `crypto` module in a short `-p` one-liner:

```
$ node -p "crypto.randomBytes(16).toString('hex')"
```

Pretty cool, isn't it?

> Note how the `crypto` module is available to the `-p` option without needing to require it (just like in the REPL mode).

How about a command to count the words in any file? This one will help us understand how to use arguments with the `node` command:

```
$ node -p "fs.readFileSync(process.argv[1])
           .toString().split(/\s+/).length" ~/.bashrc
```

Don't panic. There's a lot going on with this one. It leverages the powers of both Node and JavaScript. Go ahead and try it first. You can replace ~/.*bashrc* with a path to any file on your system.

Let's decipher this one a bit. The `readFileSync` function is part of the built-in `node:fs` module. It takes a file path as an argument and synchronously returns a binary representation of that file's data. That's why I chained a `.toString` call to it, to get the file's actual content (in UTF-8). Furthermore, instead of hardcoding the file path in the command, I put the path as the first argument to the `node` command itself and used `process.argv[1]` to read the value of that argument (see the explanation of that in the next sidebar). This enables us to use the word-counting one-liner with any file. We can alias it (without the path argument) and then use the alias with a path argument, as shown in Figure 2-1:

```
$ alias count-words="node -p 'fs.readFileSync(process.argv[1])
           .toString().split(/\s+/).length'"
```

Then once I have the content of the file, I use JavaScript's `split` method (which is available on any string) to split the content using the `/\s+/` regular expression (which means one or more spaces). This produces an array of words, and we can then count the array items with a `.length` call to find the number of words.

```
  TERMINAL                                                    bash + ⌄ ▢ 🗑 ⋯

⊗ $ count-words
  bash: count-words: command not found
○ $
● $ alias count-words="node -p 'fs.readFileSync(process.argv[1])
  >                           .toString().split(/\s+/).length'"
○ $
● $ count-words ~/.bashrc
  36
○ $
○ $ |
```

Figure 2-1. Aliasing a Node print one-liner

The process.argv Array

We know from the usage syntax that the node command can take arguments. These arguments can be any list of strings, and when you specify them, you make them available to the Node process.

The word-counting one-liner used process.argv[1]. The process object is a global scope object, and it simply represents Node's interface to the actual OS process that executes the node command. The argv property is an array that holds all the arguments you pass to the node command (regardless of how you're using the command). To understand that, run the following command:

```
$ node -p "process.argv" hello world
```

This will output the entire array of arguments. Node uses the first element in that array for the path of the node command itself, then the arguments are listed in order. That's why in the word-counting one-liner, I used the second element of argv.

Note that if you're executing a script, the path for that script will be the second element of process.argv, and the arguments (if any) will be listed starting with the third element.

The --require option (or -r) allows you to require a module before executing the main script. This is useful if you need to load a specific module before running your code or if you want to set up certain configurations or load some variable values. This one works only with CommonJS modules. For ES modules, you can use the --import option.

For example, let's say you have a Node project that requires the use of a module called dotenv, which loads environment variables from a file. Normally, you would need to include something like `require('dotenv').config()` at the beginning of your main file to use the dotenv module. However, with the -r option, you can load the module automatically without having to add it to any file:

```
$ node -r dotenv/config index.js
```

> Node supports loading environment variables from a file directly with the --env-file option. We'll see an example of that shortly.

The --watch option allows you to watch a file (and its dependencies) for changes. It automatically restarts Node when a change is detected. This is very useful in development environments. You can test it with any of the files we wrote so far. For example, to run the basic web server example from Chapter 1 in watch mode, you can use the following command:

```
$ node --watch index.js
```

This will start the server in watch mode. Make a change to the *server.js* file (change the "Hello World" string, for example), and notice how the node command will automatically restart.

The --test option makes Node look for and execute code that's written for testing. Node uses simple naming conventions for that. For example, it'll look for any files named with a *.test.js* suffix, or files whose names begin with *test-*.

There are a lot more options, but most of them are for advanced use. It's good to be aware of them so that in the future you can look up if there's one particular option that might make a task you're doing simpler.

Since Node is a wrapper around V8, and V8 itself has CLI options, the node command accepts many V8 options as well. The list of all the V8 options you can use with the node command can be printed with the following:

```
$ node --v8-options | less
```

You can set JavaScript harmony flags (to turn on/off experimental features); you can also set tracing flags, customize the engine memory management, and enable or disable many other features. As with the node command options, it's good to know that all these options exist.

Environment Variables

Toward the end of the `node -h` output, you can see a list of environment variables like `NODE_DEBUG`, `NODE_PATH`, and many more. Environment variables are another way to customize the behavior of Node or make custom data available to the Node process (similar to command arguments).

Every time you run the `node` command, you start an OS process. In Linux, the command `ps` can be used to list all running processes. If you run the `ps` command while a Node process is running (like the basic web server example), one of the listed processes will be Node (and you can see its process ID and stop it from the terminal if you need to). Here's a command to output all process details and filter the output for processes that have the word *node* in them:

```
$ ps -ef | grep "node"
```

The `process` object represents a bridge between the Node environment and the OS environment. We can use it to exchange information between Node and the OS. In fact, when you `console.log` a message, under the hood, the code is basically using the `process` object to write a string to the OS standard output (stdout) data stream.

Environment variables are one way to pass information from the OS environment (used to execute the `node` command) to the Node environment, and we can read their values using the `env` property of the `process` object.

Here's an example to demonstrate that:

```
$ NAME="Reader" node -p "'Hello ' + process.env.NAME"
```

This will output `Hello Reader`, as shown in Figure 2-2. It sets an environment variable, `NAME`, then reads its value with `process.env.NAME`. You can set multiple environment variables if you need to, either directly from the command line like this example, or using the Linux `export` command prior to executing the `node` command:

```
$ export GREETING="Hello"; export NAME="Reader"; \
  node -p "process.env.GREETING + ' ' + process.env.NAME"
```

> In Linux (and macOS), you can use a semicolon to execute multiple commands on the same line and \ to split a command into multiple lines.

```
  TERMINAL                                        🐚 bash  +  ∨  ⬚  🗑  …

● $ NAME="Reader" node -p "'Hello ' + process.env.NAME"
  Hello Reader
○ $
● $ export GREETING="Hello"; export NAME="Reader"; \
> node -p "process.env.GREETING + ' ' + process.env.NAME"
  Hello Reader
○ $
○ $ |
```

Figure 2-2. Using environment variables in Node

You can use environment variables to make your code customizable on different machines or environments. For example, the basic web server example in Chapter 1 hardcoded the port to be 3000. However, on a different machine 3000 might not be available, or you might need to run the server on a different port in a production environment. To do that, you can modify the code to use `process.env.PORT ?? 3000` instead of just `3000` (in the `listen` method) and then run the `node` command with a custom port when you need to:

```
$ PORT=4000 node index.js
```

Note that if you don't specify a port, the default port would be 3000 because I used the `??` (nullish) operator to specify a value when `process.env.PORT` does not have one. This is a common practice.

> You can't use Node's `process.env` object to change an OS environment variable. It's basically a copy of all the environment variables available to the process.

The list of environment variables shown toward the end of the `node -h` output are Node's built-in environment variables. These are variables that Node will look for and use if they have values. Here are a few examples:

NODE_PATH
This can be used to simplify import statements by using absolute paths instead of relative ones.

NODE_OPTIONS
This is an alternative way to specify the options Node supports instead of passing them to the command line each time.

NODE_DEBUG

This can be used to tell Node to output more debugging information when it uses certain libraries (which we specify as a comma-separated list). For example, with NODE_DEBUG=fs,http, Node will start outputting debugging messages when the code uses either the node:fs or node:http modules. Many packages support this environment variable. Figure 2-3 shows an example of using it with the node:http module.

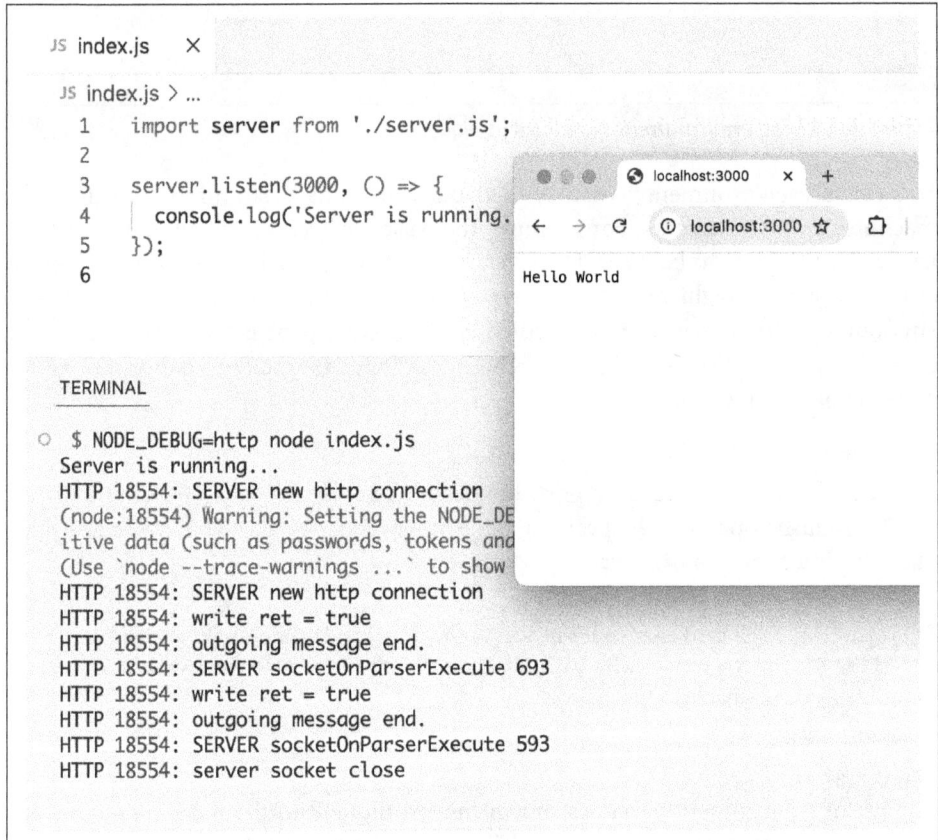

```
JS index.js    ×

JS index.js > ...
  1    import server from './server.js';
  2
  3    server.listen(3000, () => {
  4      console.log('Server is running.
  5    });
  6
```

localhost:3000

← → C ⓘ localhost:3000 ☆

Hello World

```
TERMINAL

$ NODE_DEBUG=http node index.js
Server is running...
HTTP 18554: SERVER new http connection
(node:18554) Warning: Setting the NODE_DE
itive data (such as passwords, tokens and
(Use `node --trace-warnings ...` to show
HTTP 18554: SERVER new http connection
HTTP 18554: write ret = true
HTTP 18554: outgoing message end.
HTTP 18554: SERVER socketOnParserExecute 693
HTTP 18554: write ret = true
HTTP 18554: outgoing message end.
HTTP 18554: SERVER socketOnParserExecute 593
HTTP 18554: server socket close
```

Figure 2-3. Using NODE_DEBUG with the http module

You can also put all the environment variables you need to set in a file (like an *.env* file for example) and then instruct Node to include all of the values defined in that file in the `process.env` object, using the `--env-file` option of the `node` command. Say, for example, you have the following *.env* file:

```
PORT=3000
NODE_DEBUG=fs,http
```

You can execute a Node script with these environment variables set using the following command:

```
$ node --env-file=.env script.js
```

> You can use multiple environment files if you need to.

REPL Mode

In Node's REPL mode, as we learned in Chapter 1, you can type any JavaScript code, and Node will execute it and automatically print its result. This is a convenient way to quickly test short JavaScript expressions (and it works for bigger code too). There are a few other helpful things you can do in REPL mode beyond the quick tests.

In REPL mode, you usually type an expression (for example, `0.1 + 0.2`) and hit Enter to see its result. You can also type statements that are not expressions, such as `let v = 21;`. In this case, when you hit Enter the variable v will be defined, and the REPL mode will print `undefined` since that statement does not evaluate to anything. If you need to clear the screen, you can do so with Ctrl + L.

If you try to define a function, you can write the first line and hit Enter; the REPL mode will detect that your line is not complete, and it will go into a *multiline* mode so that you can complete it, as shown in Figure 2-4. Try to define a small function to test that.

```
  TERMINAL                                           node  + v  []  🗑  ...

○  $ node
   >
   > function add(x, y) {
   ... return x + y;
   ... }
   undefined
   >
   > add(4, 3)
   7
   >
   > |
```

Figure 2-4. Node REPL multiline mode

The REPL multiline mode is limited, but there's an integrated basic editor available within REPL sessions as well. While in a REPL session, type .editor to start the basic editor mode. Then you can type as many lines of code as you need; you can define multiple functions or paste code from the clipboard. When you're done, hit Ctrl + D to have Node execute all the code you typed in the editor.

The .editor command is one of many REPL commands that you can see by typing the .help command:

```
> .help
.break    Sometimes you get stuck, this gets you out
.clear    Alias for .break
.editor   Enter editor mode
.exit     Exit the REPL
.help     Print this help message
.load     Load JS from a file into the REPL session
.save     Save all evaluated commands in this REPL session to a file

Press Ctrl+C to abort current expression, Ctrl+D to exit the REPL
```

The .break command lets you get out of weird cases in REPL sessions. For example, when you paste some code in Node's multiline mode and you're not sure how many curly braces you need to get to an executable state, you can completely discard your pasted code by using a .break command (or pressing Ctrl + C once). This saves you from killing the whole session to get yourself out of situations like these.

The .exit command exits the REPL session (just like Ctrl + D).

The `.save` command enables you to save all the code you typed in one REPL session into a file. The `.load` command enables you to load JavaScript code from a file and make it all available within the REPL session. Both of these commands take a filename as an argument.

One of my favorite things about Node's REPL mode is how I can inspect basically everything that's available natively in Node without needing to require it. All the built-in modules (like `fs`, `http`, etc.) are preloaded in a REPL session and you can use the Tab key to inspect their APIs.

Just like in a terminal or editor, hitting the Tab key once in a REPL session will attempt to autocomplete anything you partially type. Try typing `cr` and hitting Tab to see it get autocompleted to `crypto`. Hitting the Tab key twice can be used to see a list of all the possible things you can type from whatever partially typed text you have. For example, type `a` and hit Tab twice to see all the available global scope objects that begin with `a`, as shown in Figure 2-5.

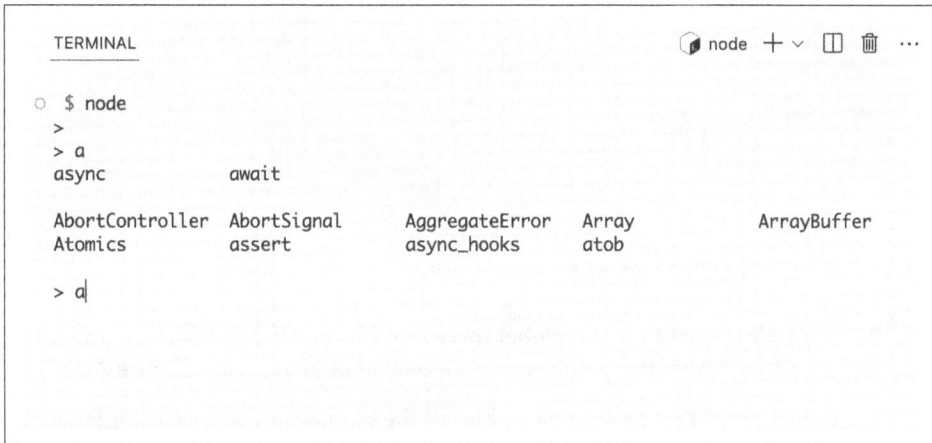

```
TERMINAL                                              node + ∨  □  🗑  ⋯

○  $ node
   >
   > a
   async            await

   AbortController  AbortSignal    AggregateError   Array      ArrayBuffer
   Atomics          assert         async_hooks      atob

   > a
```

Figure 2-5. Node REPL autocomplete

This is great if you need to type less and avoid typing mistakes, but it gets better. You can use the Tab key to inspect the methods and properties available on any object. For example, type `Array.` and hit Tab twice to see all the methods and properties that you can use with the JavaScript `Array` class. This works with Node modules as well. Try it with `fs.` or `http.`.

It even works with objects that you create. For example, create an empty array using `let myArr = [];`, then type `myArr.` and hit Tab twice to see all the methods available on an array instance, as shown in Figure 2-6.

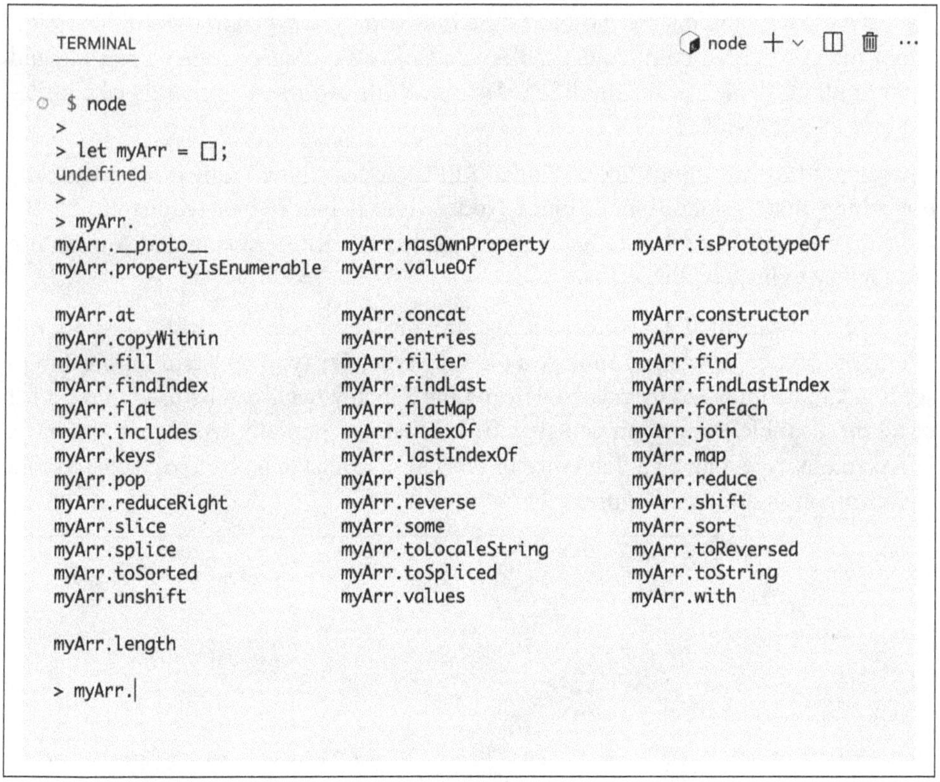

```
  TERMINAL                                                    node  + ∨  ⬚  🗑  ⋯

  ○  $ node
     >
     > let myArr = [];
     undefined
     >
     > myArr.
     myArr.__proto__              myArr.hasOwnProperty        myArr.isPrototypeOf
     myArr.propertyIsEnumerable   myArr.valueOf

     myArr.at                     myArr.concat                myArr.constructor
     myArr.copyWithin             myArr.entries               myArr.every
     myArr.fill                   myArr.filter                myArr.find
     myArr.findIndex              myArr.findLast              myArr.findLastIndex
     myArr.flat                   myArr.flatMap               myArr.forEach
     myArr.includes               myArr.indexOf               myArr.join
     myArr.keys                   myArr.lastIndexOf           myArr.map
     myArr.pop                    myArr.push                  myArr.reduce
     myArr.reduceRight            myArr.reverse               myArr.shift
     myArr.slice                  myArr.some                  myArr.sort
     myArr.splice                 myArr.toLocaleString        myArr.toReversed
     myArr.toSorted               myArr.toSpliced             myArr.toString
     myArr.unshift                myArr.values                myArr.with

     myArr.length

     > myArr.|
```

Figure 2-6. Exploring methods with autocomplete

Tab discoverability works on the global level too. If you hit Tab twice on an empty line, you get a list of everything that is globally available, as shown in Figure 2-7.

This is a big list, but it's a useful one. It has all the globals in the JavaScript language itself (like `Array`, `Number`, `Math`, etc.), and it has all the globals from Node (like `process`, `setTimeout`, etc.), and it also lists all the core modules that are available natively in Node (like `fs`, `http`, etc.).

```
$ node
>
>
AbortController                      AbortSignal
AggregateError                      Array
ArrayBuffer                         Atomics
BigInt                              BigInt64Array
BigUint64Array                      Blob
Boolean                             BroadcastChannel
Buffer                              ByteLengthQueuingStrategy
CompressionStream                   CountQueuingStrategy
Crypto                              CryptoKey
CustomEvent                         DOMException
DataView                            Date
DecompressionStream                 Error
EvalError                           Event
EventTarget                         File
FinalizationRegistry                Float32Array
Float64Array                        FormData
Function                            Headers
Infinity                            Int16Array
Int32Array                          Int8Array
Intl                                Iterator
JSON                                Map
Math                                MessageChannel
MessageEvent                        MessagePort
NaN                                 Navigator
Number                              Object
Ponfonmanco                         PonfonmancoEntnv
```

Figure 2-7. Hitting Tab twice on an empty line

In the list of all global things, you'll notice an underscore character (_). This is a handy REPL session variable that stores the value of the last evaluated expression. For example, after executing a `Math.random()` line, you can type _ to access that same random value. You can even use it in any place where you use a JavaScript expression. Try `let random = _;`.

You can use the `node:repl` module to create your own custom REPL server. You can customize many things, like the prompt, the input and output streams, whether to use colors or not, and a few more options. You can also attach your own global context objects to it.

Here's a custom REPL example that'll start a REPL session with a different prompt, in strict mode, and it'll not output the return value if it's `undefined`. It'll also make the `lodash` library available globally in your custom REPL sessions (see Figure 2-8):

```js
import { start, REPL_MODE_STRICT } from 'repl';
import lodash from 'lodash';

const replServer = start({
  prompt: '... ',
  ignoreUndefined: true,
  replMode: REPL_MODE_STRICT,
});

replServer.context.lodash = lodash;
```

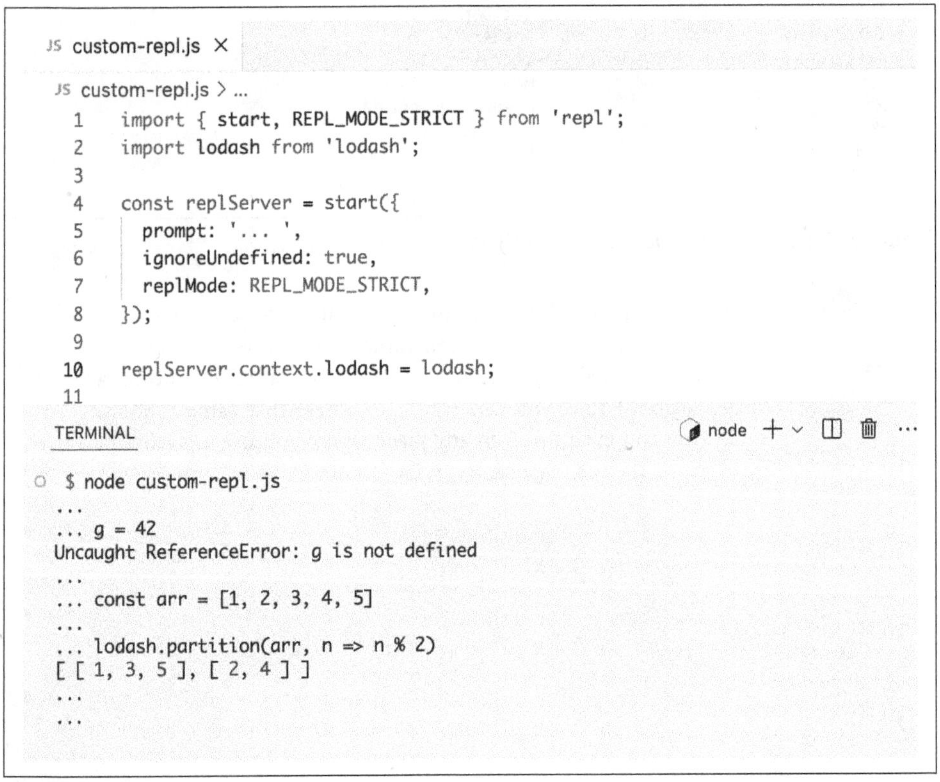

Figure 2-8. Using a custom REPL server

Node Modules

The word *module* means a reusable piece of code. Something you can include and use in any application, as many times as you need.

In Node, the word *script* is usually used for a piece of code that's executed once with the node command. Any other files or folders that are required or imported are what's referred to as modules.

When you specify a module as a dependency, Node goes through a few key steps to complete the module loading process: resolution and reading of the module contents, isolating the module scope, executing the module code, and caching the module.

Resolving Modules

Node uses the following procedure to determine how to find a module that is being imported.

If the module name does not start with a . (denoting a relative path) or a / (denoting an absolute path), Node will first check if the module is a built-in one. If it is, it'll load and execute it directly.

If the module is not built-in, Node will look for it under *node_modules* folders starting from the location where the importing module is and going up in the folders hierarchy. For example, if the importing module is in */User/samer/efficient-node/src*, Node will first look under *src* for a *node_modules* folder. If it does not find one, it'll look next under *efficient-node*, and so on, all the way to the root path.

You can use this lookup procedure to localize module dependencies by having multiple *node_modules* folders in your project, but that generally increases the complexity of the project. You can also use this lookup procedure to have multiple projects share a *node_modules* folder by placing that folder in a parent folder common to all projects, or even have a global *node_modules* folder for all projects on one server. While this might be useful in some cases, having a single *node_modules* folder per project is the standard and recommended practice.

If the imported module starts with a . or /, Node will look for it in the folder or file specified by the relative or absolute path.

> For CommonJS modules, if you set the NODE_PATH environment variable before executing a script, Node will first look for any required modules in the paths specified by NODE_PATH (which can be a single path or multiple paths separated by a comma). This can be useful to use short absolute paths instead of confusing relative ones.

If you need to only resolve the module and not execute it, you can use the `require.resolve()` function for CommonJS modules, or the `import.meta.resolve()` function for ES modules. These functions do not load the module. They just verify that it exists, and they'll throw an error if it does not.

Loading Modules

Once the path of a module is resolved successfully, Node will read the content of the module and determine its type.

A module can be a CommonJS module or an ES module. Supported file extensions are *.js*, *.cjs*, and *.mjs*. It can be a single file or a directory with a *package.json* that specifies what files in the directory can be imported.

A module can also be a JSON file (*.json* extension). When you import a JSON file, you get a JavaScript object representing the data in that JSON file:

```
// In CommonJS modules:
const data = require('./file.json');

// In ES modules with static import:
import data from './file.json'
  with { type : 'json' };

// In ES modules with dynamic import:
const { default: data } = await import('./file.json', {
  with: { type: 'json' },
});
```

> The `with` keyword in this example is used to specify the `type` import attribute. The *import attributes* feature gives the runtime instructions on how a module should be loaded. It's a security standard to prevent executing malicious code. It can be used with other module types as well (for example, a "css" module in a browser).

A module can also be a Node add-on compiled file. Node add-ons are dynamically linked objects implemented in a low-level language like C or C++ and compiled to be loaded as ordinary Node modules. Node has an API known as *Node-API* that's dedicated to building native add-ons. It's independent from the underlying JavaScript runtime. If you need a module with high performance, or you need it to access system resources or integrate with C or C++ libraries, you can use Node-API to build an add-on and use it as you would use any other built-in Node module.

Add-ons are not supported with ES module imports. They can instead be loaded using the `module.createRequire()` function.

Scoping Modules

JavaScript functions can be called with any number of arguments. The `arguments` keyword can be used to access the list of all arguments a function is called with (see Figure 2-9).

```
JS arguments.js ✕

JS arguments.js > ...
1    function args() {
2        console.log(arguments);
3    }
4
5    args(1, 5, 9, 11);
6

TERMINAL                                                    bash + ∨  ⬛ 🗑 ⋯

● $ node arguments.js
  [Arguments] { '0': 1, '1': 5, '2': 9, '3': 11 }
○ $ |
```

Figure 2-9. The implicit `arguments` object

If you need to have a function with a dynamic number of arguments, you should use explicit *rest parameters* instead of the implicit `arguments` keyword.

Node wraps all CommonJS modules with a function to give them a private scope. That wrapping function is called with five implicit arguments. To see that in action, print the value of the `arguments` keyword in the top-level scope of a CommonJS module, as shown in Figure 2-10.

```
JS module.cjs  ✕

JS module.cjs > [e] id
  1    exports.id = '123';
  2    💡
  3    console.dir(arguments, { depth: 1 });
  4
```

```
TERMINAL                                           🎲 bash  + ∨  ⬚  🗑  ⋯

● $ node module.cjs
  [Arguments] {
    '0': { id: '123' },
    '1': [Function: require] {
      resolve: [Function],
      main: [Object],
      extensions: [Object: null prototype],
      cache: [Object: null prototype]
    },
    '2': {
      id: '.',
      path: '/Users/samer/efficient-node',
      exports: [Object],
      filename: '/Users/samer/efficient-node/module.cjs',
      loaded: false,
      children: [],
      paths: [Array],
      [Symbol(kIsMainSymbol)]: true,
      [Symbol(kIsCachedByESMLoader)]: false,
      [Symbol(kIsExecuting)]: true
    },
    '3': '/Users/samer/efficient-node/module.cjs',
    '4': '/Users/samer/efficient-node'
  }
○ $ |
```

Figure 2-10. CommonJS module wrapping

These five implicit arguments are (in order): exports, require, module, __filename, and __dirname. When you use these within a CommonJS module, you are not using a global variable; you're using an argument from the implicit wrapping function.

The exports, require, and module arguments are Node's way to manage a CommonJS module's API and its dependencies. The __filename value has the full path of the module file. The __dirname value has the path to the directory where the module file is located.

Similar to CommonJS module wrapping, ES modules are executed in an implicit scope, but there is no wrapping function, and the five implicit arguments are not defined at all. Instead, an ES module API and dependencies are managed with import/export statements.

If you need to access the filename or directory name of an ES module, you can use `import.meta.filename` and `import.meta.dirname`.

With this scoping in modules, all the variables you define in a module are local to that module. If you need to define a global variable, you can use the global scope object `globalThis`. Any properties you add to that object become global variables. It's good to know that you can do that, but you should avoid using global variables as they can be problematic for many reasons.

Executing Modules

This is the step where Node will execute the code in a module and finalize its dependencies and exports.

One common coding practice is to put any configurable variables that are used to seed or run an application into their own modules. An example of such configurable variables are the `PORT` and `HOST` on which a web server will run.

Let's create a *config.cjs* file to host these two configurable variables. The *.cjs* extension makes it a CommonJS module that will be wrapped for scoping. This module will have the five implicit arguments.

> I'll provide the equivalent ES module syntax below the CommonJS module syntax. You can use the *.mjs* extension if you want to test the examples with ES modules.

As shown in Figure 2-11, the `exports` argument will start out as an empty object.

```
JS config.cjs  ×

  JS config.cjs
   1    console.log(exports);
   2
```

```
TERMINAL                                          bash  + ∨  ⊡  🗑  ⋯
● $ node config.cjs
  {}
○ $ |
```

Figure 2-11. The `exports` *argument*

To define the API of the *config.cjs* module, we just define properties on the exports object. Properties can be static values or any type of object in JavaScript (like a function, a class, or a promise):

```
console.log('Loading config.cjs');

exports.PORT = process.env.PORT ?? 3000;
exports.HOST = process.env.HOST ?? 'localhost';
exports.SERVER_URL = (
  protocol = process.env.PROTOCOL ?? 'http',
) => `${protocol}://${exports.host}:${exports.PORT}`;

// In ES modules
export const PORT = process.env.PORT ?? 3000;
export const HOST = process.env.HOST ?? 'localhost';
export const SERVER_URL = (
  protocol = process.env.PROTOCOL ?? 'http',
) => `${protocol}://${exports.host}:${exports.PORT}`;
```

Note how I used process.env variables to make the configurations customizable on different environments. I also made SERVER_URL a function that receives a protocol argument, which is customizable through the environment as well. Making a configuration value a function allows it to be customizable at runtime.

When we require this *config.cjs* module in another module, the require function call returns the exports object. Let's test that in an *index.cjs* file:

```
const config = require('./config.cjs');
console.log(config);

// Or we can use destructuring
// const { PORT, HOST } = require('./config.cjs');

// In ES modules
import * as config from './config.mjs';
console.log(config);

// Or we can use named imports
// import { PORT, HOST } from './config.cjs';
```

Figure 2-12 shows the output.

Now we can say that the *index.cjs* module *depends* on the *config.cjs* module. This is where the term *dependency management* comes from. We are managing the dependencies of a module here and bringing one module's API to use in another module.

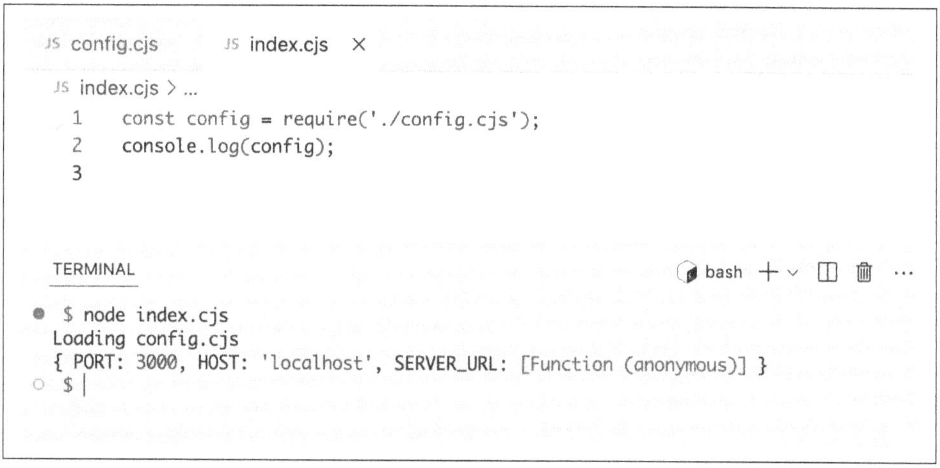

```
JS config.cjs        JS index.cjs    ✕

JS index.cjs > ...
  1     const config = require('./config.cjs');
  2     console.log(config);
  3


TERMINAL                                                     bash  + ∨  □  🗑  ⋯

●  $ node index.cjs
   Loading config.cjs
   { PORT: 3000, HOST: 'localhost', SERVER_URL: [Function (anonymous)] }
○  $ |
```

Figure 2-12. Requiring a CommonJS module

The `exports` argument in CommonJS modules is actually an alias to `module`
`.exports`. The latter is what's returned when we invoke the `require` function. In
some cases, you might need the top-level API object to be a function or a class, or
anything else that's not a simple aliased object. In these cases, you'll need to change
the value of `module.exports` itself to define your special API.

For example, let's say that we want all the configuration properties to be the result of
executing a function. This might be helpful for testing as we can mock the configuration function differently for different tests. To make the top-level API object a function, you need to use `module.exports`. Here's an example of how we can do that for
config.cjs:

```
module.exports = () => {
  return {
    PORT: process.env.PORT ?? 3000,
    HOST: process.env.HOST ?? 'localhost',
    SERVER_URL: (protocol = process.env.PROTOCOL ?? 'http') =>
      `${protocol}://${exports.host}:${exports.PORT}`,
  };
};

// In ES modules
export default () => {
  return {
    PORT: process.env.PORT ?? 3000,
    HOST: process.env.HOST ?? 'localhost',
    SERVER_URL: (protocol = process.env.PROTOCOL ?? 'http') =>
      `${protocol}://${exports.host}:${exports.PORT}`,
  };
};
```

With that, to use the configuration value in *index.cjs*, we'll need to invoke what we get from the `require` function:

```
const config = require('./config.cjs');

console.log(
  config(), // Note how we are invoking this
);
// In ES modules:
import config from './config.mjs';

console.log(
  config(), // Note how we are invoking this
);
```

This method is often helpful when you need to use the *dependency injection* design pattern, which is when some modules are injected into other modules to create more flexibility and make modules more reusable.

If you need to make a Node module executable from the CLI as a script, you can use the `require.main` property to check if the module is being run directly. The `require.main` value will equal the `module` argument in that case. Figure 2-13 has an example of a simple module using that check to determine what to do.

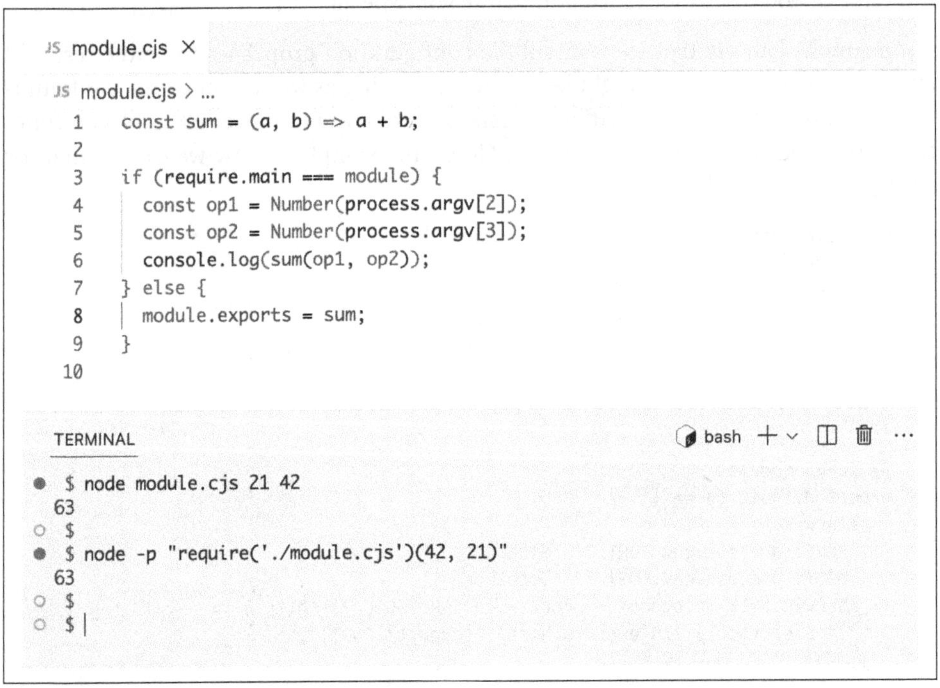

```
JS module.cjs  ✕

JS module.cjs > ...
 1    const sum = (a, b) => a + b;
 2
 3    if (require.main === module) {
 4      const op1 = Number(process.argv[2]);
 5      const op2 = Number(process.argv[3]);
 6      console.log(sum(op1, op2));
 7    } else {
 8      module.exports = sum;
 9    }
10
```

```
TERMINAL                                              🍥 bash  + ∨  ⬚  🗑  ⋯

● $ node module.cjs 21 42
  63
○ $
● $ node -p "require('./module.cjs')(42, 21)"
  63
○ $
○ $ |
```

Figure 2-13. The `require.main` check

ES modules have no equivalent simple check to determine if they are run directly, but `import.meta.url` can be used along with `process.argv` to do a similar check. The `es-main` npm package has a good implementation of that.

Caching Modules

To understand another concept about how Node modules work, let's repeat the `require` line in *index.cjs* multiple times:

```
require('./config.cjs');
require('./config.cjs');
require('./config.cjs');
```

Given these three `require` lines, when we execute *index.cjs*, how many times will the "Loading config.cjs" line in *config.cjs* be outputted?

The answer is not three times. It'll be outputted only once, as shown in Figure 2-14.

Figure 2-14. Node module caching

Both CommonJS modules and ES modules in Node are cached after the first call. A module is executed the first time you import it; then when you import it again, Node loads it up from a cache.

If you look at a frontend application like React, for example, all component files import the `React` module, and that's OK, because only the first import will do the work; the rest will use the cache.

But what if you do want the `console.log` message to show up every time we require *config.cjs*?

You can make the top export of *config.cjs* a function instead of an object, put all the code there inside the function, and call that function every time you need the code to be executed. The cache, in that case, will cache the definition of the function.

Summary

Node CLI has many powerful options that we can control. We can pass arguments to it and set environment variables before running it. Both of these options allow us to pass data from the OS environment to a running Node process. Node's `process` object is the bridge.

Node's REPL mode is a good way to test simple expressions, explore everything you can use in Node, and take a quick look at the API of anything, including core modules, installed modules, and even objects you instantiate.

CommonJS modules in Node are implicitly wrapped in a function and are passed five arguments. ES modules have a private scope as well.

We use the `exports` object in CommonJS modules or `export` statements in ES modules to define the API of a module. Modules that need to depend on other modules use the `require` function or `import` statements to access a dependency API.

Node manages a cache for all modules. To discover where a module is, Node follows a predefined set of rules depending on the path of the module. A path can be a relative one, an absolute one, or just a name. For the latter case, Node looks for the module in *node_modules* folders.

In the next chapter, we'll do a deep dive into how Node handles asynchronous operations and learn about the event-driven nature of Node modules.

CHAPTER 3

Asynchrony and Events

We learned that the most important concept in Node is its event-driven model, which is also known as the *non-blocking model*. Node associates asynchronous operations with events and internally takes care of running them independently from other operations. Once an asynchronous operation is done running, Node schedules the execution of your code that depends on that operation.

In this chapter, we'll expand on the important concepts of events and their handler functions as callbacks, promises, and listeners. We'll learn how Node manages the scheduling of asynchronous operations with its event loop and event queues, and see examples of how to create and use custom event emitters and listener functions.

Sync Versus Async Handling

A web browser is a single-threaded environment where all user-related code runs in one main thread. If you have a slow function in your website, users will not be able to even scroll as long as that function is executing. That's because scrolling needs to use the single busy thread.

You get a single thread for your Node code too. Node works with other threads internally, and you can also manually create your own threads when you need to, but it's extremely important that you don't do anything to block your main thread. We saw simple abstract examples of that in Chapter 1. Let's look into an example where blocking the main thread means an entire web server will not be able to serve requests.

If you have a slow synchronous operation in your Node's web server, an incoming HTTP request will block the single precious thread, and your entire web server will not be able to handle any other incoming requests while the slow operation is running under the first incoming request.

To see that in action, let's use the long-running for loop to simulate a slow operation within the simple web server example from Chapter 1. To demonstrate the effects of blocking the main thread, let's run the slow operation for only the first incoming request to the server, and skip it for all subsequent incoming requests. We can do that with a simple counter variable:

```
import http from 'node:http';

const slowOperation = () => {
  for (let i = 0; i < 1e9; i++) {
    // Simulate a synchronous delay
  }
};

let counter = 0;

const server = http.createServer((req, res) => {
  counter = counter + 1;
  if (counter === 1) {
    slowOperation();
    res.end('Slow Response');
  } else {
    res.end('Normal Response');
  }
});

server.listen(3000, () => {
  console.log('Server is running...');
});
```

Note how the first request (when counter is 1) will have to wait on the slow operation to complete. The other requests do not need the slow operation and should be fast. However, because the slow operation is synchronous, not only is the first request going to be slow, but all subsequent requests will have to wait on the first one as well!

To test that, run this example and then connect to *http ://localhost:3000* multiple times, as shown in Figure 3-1.

```
JS server-sync.js ✕

JS server-sync.js > [ø] slowOperation
   3    const slowOperation = () => {
   4      for (let i = 0; i < 1e10; i++) {
   5        // Simulate a synchronous delay
   6      }
   7    };
   8
   9    let counter = 0;
  10
  11    const server = http.createServer((req, res) => {
  12      counter = counter + 1;
  13      if (counter === 1) {
  14        slowOperation();
  15        res.end('Slow Response');
  16      } else {
  17        res.end('Normal Response');
  18      }
```

TERMINAL

```
○ $ node server-sync.js
  Server is running...
```

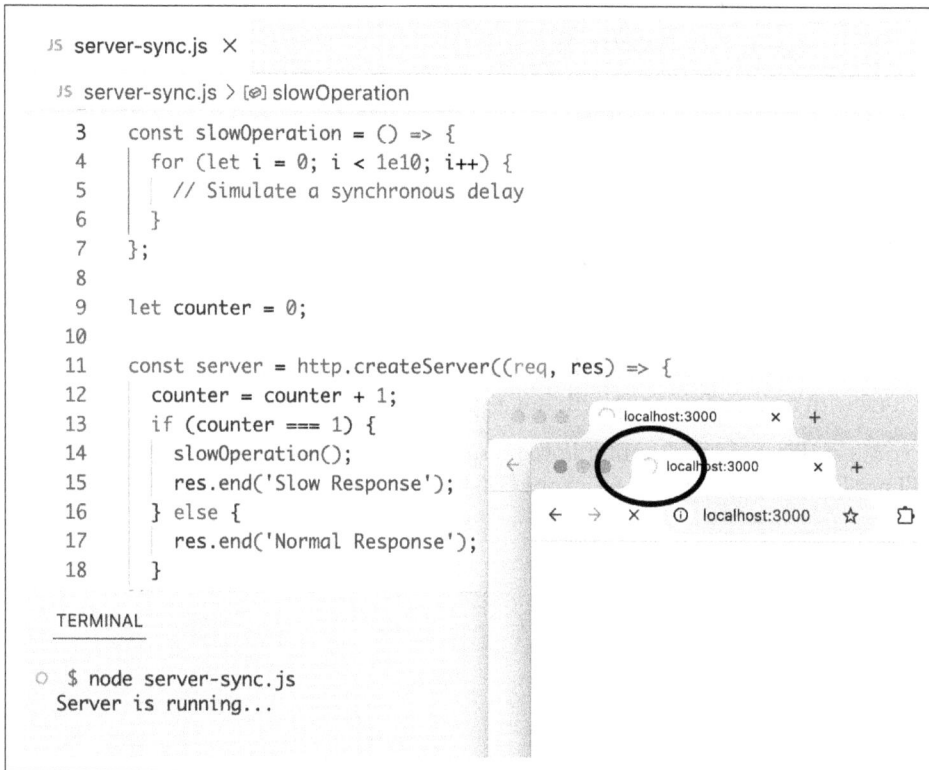

Figure 3-1. A synchronous slow operation

Note how the second browser is spinning, waiting on the first one to be done. Obviously this is very bad, and you should never do something like that.

To simulate executing a slow operation without blocking the main thread, we can use the setTimeout function:

```
import http from 'node:http';

const slowOperation = (cb) => {
  // Simulate an asynchronous delay
  setTimeout(
    () => cb(),
    15_000, // delay is in ms
  );
};

let counter = 0;

const server = http.createServer((req, res) => {
  counter = counter + 1;
  if (counter === 1) {
```

```
      slowOperation(() => {
        res.end('Slow Response');
      });
    } else {
      res.end('Normal Response');
    }
  });

  server.listen(3000, () => {
    console.log('Server is running...');
  });
```

The gist of these examples is the same. The first request is delayed because it needs to run after a slow operation. The difference in this example is that the simulated delay is done with an asynchronous API, not a synchronous, long `for` loop.

If you test this code with multiple requests, only the first one will be delayed, but it will not block any other requests from finishing fast, as shown in Figure 3-2.

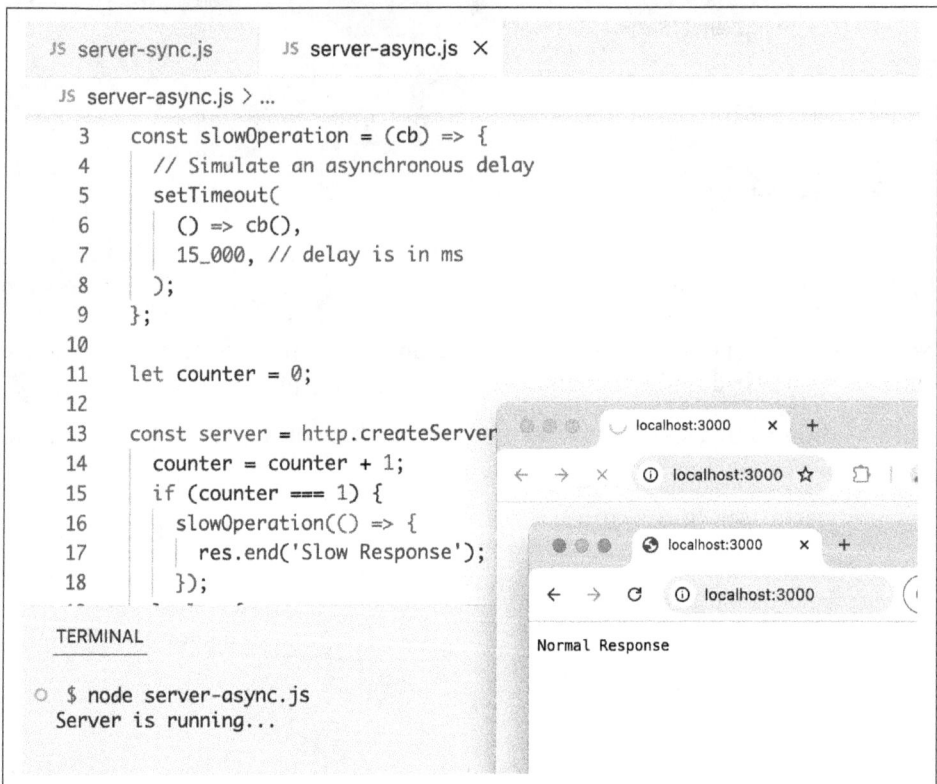

Figure 3-2. An asynchronous slow operation

Note how the second browser in my test responded immediately while the first one was still waiting for the slow operation. This server is basically doing multiple tasks in parallel now, and we did not need to deal with any threads to make it do so.

> There will be cases where your slow operation cannot be made asynchronous with built-in APIs. I'll cover an example of how to handle that without blocking the main thread in Chapter 7.

Handler Functions

Let's review a typical example of an asynchronous callback-style function and how it can be used:

```
// Callback Style

const readFileAsArray = function(file, cb) {
  fs.readFile(file, function(err, data) {
    if (err) {
      return cb(err);
    }
    const lines = data.toString().trim().split('\n');
    cb(null, lines);
  });
};
```

The readFileAsArray function takes a file path and a callback function. It reads the file content, splits it into an array of lines, and invokes the callback function after that. It passes the array of lines as an argument to the callback function. If an error occurs while reading the content of the file, the callback function will be invoked with that error as its first argument.

> The readFileAsArray function is asynchronous because it uses the asynchronous API method fs.readFile. A function becomes asynchronous if it uses any asynchronous functions or if it returns a Promise object. We can also make a function asynchronous by using the async keyword.

Assume that we have the file *numbers.txt* in the same directory with content like this:

```
10
11
12
13
14
15
```

If we have a task to count the odd numbers in that file, we can use the `readFileAs` `Array` function to simplify that task:

```
readFileAsArray('./numbers.txt', (err, lines) => {
  if (err) throw err;
  const numbers = lines.map(Number);
  const oddNumbers = numbers.filter(n => n%2 === 1);
  console.log('Odd numbers count:', oddNumbers.length);
});
```

The `readFileAsArray` produces an array of lines in a file. For *numbers.txt*, that would be an array of strings that represent numbers. The callback function in this example parses these strings as numbers and then counts the odd numbers.

Node's callback style is used here. The standard way to structure this style is to have the callback function as the last argument for the asynchronous function. This is one very common convention about Node's callback style. The other convention is how callback functions always have an error-first argument (named `err` in this example). Node's callback style is sometimes referred to as the *error-first* callback style.

Beyond consistency, the error-first callback style conventions are in place for readability and better error handling. If you need to work with Node callback style, you should always structure your functions with these conventions.

If an error happens in the asynchronous function, it'll call the callback function, passing that error down via the `err` argument. If no errors happen, the value of the `err` argument will be `null`. Any callback function written for this style should always check for the presence of that `err` argument before using anything else it has access to (like `lines` in this example). If the error check is not there, errors will just be ignored, and that's bad.

There are a few problems worth mentioning here about callbacks, the most popular of which is what the community named *the pyramid of doom*, or *callback hell*. This happens when you need to do multiple asynchronous operations that depend on each other. You'll have to nest callbacks within each other, making a pyramid-shaped code. This makes it tough to understand the flow of the program.

Another important disadvantage to the callback style is how the callbacks take control of what needs to happen after an asynchronous task is done, and that basically means that we'll have to trust that they'll do the right thing. I explain this in "An Analogy for Promises" on page 69.

Synchronous Callbacks

It's important to understand that the callback pattern by itself does not mean the call-back will happen asynchronously. A function can still call its callback synchronously. For example, here's a function that accepts a callback function and may invoke that callback function either synchronously or asynchronously based on a condition:

```
// Example function with both
// synchronous and asynchronous returns

function fileSize (fileName, cb) {
  if (typeof fileName !== 'string') {
    return cb(new TypeError('argument should be string')); // Sync
  }
  fs.stat(fileName, (err, stats) => {
    if (err) { return cb(err); } // Async
    cb(null, stats.size); // Async
  });
}
```

The cb function here will be invoked synchronously if fileSize is called with a file Name that's not a string, and asynchronously otherwise. This is a bad practice because it might lead to unexpected errors. You should design your functions to consume call-backs either always synchronously or always asynchronously. For example, we could use setTimeout in the if statement to make the fileSize consistently work asynchronously.

Promises

Instead of passing a callback function as an argument to an asynchronous function, then handling both success and error cases in the same place, a Promise object allows us to handle success and error cases separately, and it also allows us to chain multiple asynchronous calls instead of having to nest them.

If the readFileAsArray function supports promises, we can use it as follows:

```
readFileAsArray('./numbers.txt')
  .then(
    (lines) => {
      const numbers = lines.map(Number);
      const oddNumbers = numbers.filter(n => n%2 === 1);
      console.log('Odd numbers count:', oddNumbers.length);
    }
  )
  .catch(
    (err) => console.error(err)
  );
```

We invoke a .then function on the return value of the asynchronous function. To support this syntax, the asynchronous function needs to return a Promise object. In JavaScript, you can create a Promise object using the Promise constructor. We'll see how shortly.

In this example, the .then function gives us access to the same lines array that we get in the callback version, and we can do our processing on it as before. To handle errors, we can chain a .catch call, and in there, we get access to the error object if an error happens.

> When working with Promise objects using this syntax, you should always add a .catch call (just like you should always check the err argument in callback functions). When using the async/await syntax, we can use the try/catch statement to handle any errors from Promise objects.

Here's how we can modify the readFileAsArray function to support a promise interface, in addition to the callback interface it already supports:

```
// Making a callback function support promises

const readFileAsArray = function (file, cb = () => {}) {
  return new Promise((resolve, reject) => {
    fs.readFile(file, function (err, data) {
      if (err) {
        reject(err);
        return cb(err);
      }
      const lines = data.toString().trim().split('\n');
      resolve(lines);
      cb(null, lines);
    });
  });
};
```

We make the function return a Promise object (using new Promise) that simply wraps the fs.readFile async call. Whenever we need to invoke the callback with an error, we invoke the promise reject function as well, and whenever we need to invoke the callback with data, we invoke the promise resolve function as well.

The only other thing we needed to do in this case is to have a default value for the cb argument in case the code is being used with the promise interface. For that, we can simply use a default empty function in the argument: () => {}.

Now the readFileAsArray function can be used with both the callback pattern and the promise pattern.

While you can still make a pyramid of doom using promises, avoiding it is easy. Promise objects support chaining. If you need to do an async operation within the first .then handler, you can chain the handler for that operation on the original call instead of nesting it. For example, say that you have a *users* resource and a *posts* resource, and your code needs to start with a userEmail value, use that in an async operation to get the userId value, then use that in another async operation to get the list of posts from the user. The code would look something like this:

```
getUserIdFromEmail(userEmail)
  .then((userId) => getUserPosts(userId))
  .then((userPosts) => {
    // Do something with userPosts
  })
  .catch((err) => {
    // Do something with err
  });
  .finally(() => {
    // Do something final
    // even if the promise is rejected
  });
```

Note how we did not need to nest these two operations, and this remains the case no matter how many more asynchronous operations we need. This is certainly more readable than what the callback style could offer for the case.

async/await

The async/await pattern improves not only the readability of the code but also its error handling. It also simplifies variable scopes around your asynchronous code.

Here's how we can consume the promisified version of the readFileAsArray function with async/await:

```
async function countOdd () {
  try {
    const lines = await readFileAsArray('./numbers.txt');
    const numbers = lines.map(Number);
    const oddCount = numbers.filter(n => n%2 === 1).length;
    console.log('Odd numbers count:', oddCount);
  } catch(err) {
    // Do something with err
  }
}

countOdd();
```

We first create an async function, which is just a regular function with the word async before it. Inside the async function, we call the readFileAsArray function as if it returns the lines variable (although it returns a promise). To make that work, we use the keyword `await` in front of the function call. After that, we continue the code as if the `readFileAsArray` call was synchronous.

To work with errors, we wrap the async call in a `try/catch` statement. To get things to run, we execute the async function.

The `async/await` syntax simplifies the code, and that is especially true when you need to invoke multiple async functions that depend on each other. Here's an example of how the `async/await` syntax simplifies the code flow of multiple async functions:

```
async function getUserInfo(userEmail) {
  try {
    const userId = await getUserIdFromEmail(userEmail);
    const userPosts = await getUserPosts(userId);

    // Do something with userPosts (and userId)
  } catch (err) {
    // Do something with err
  }
}
```

Besides not having to deal with `.then`/`.catch` calls and making sure `.then` functions return promises to be chained, there's a subtle but important advantage for the `async/await` version; we can use the results of all the sequential async calls in the same place! We'll need to write more code with the `.then`/`.catch` version to accomplish that.

> In many cases, you'll need to execute promises in parallel rather than in sequence. One way to do that is with a `Promise .all([promise1, promise2, promise3])` call. You can also use `await [promise1, promise2, promise3]`. The result of both these calls is an array containing the resolved objects in order.

We can use the `async/await` feature with any function that supports a promise interface. We can't use it with callback-style async functions, but as we've seen before, any function designed with the callback style can easily be promisified. In fact, Node has a built-in utility function that can promisify any function that follows the error-first callback style. For example, since the original callback-based `readFileAsArray` function follows the error-first callback style, we can use Node's built-in `node:util` module to promisify it:

```
import { promisify } from 'node:util';

// const readFileAsArray = function(file, cb) { ... }

const readFileAsArrayPromise = promisify(readFileAsArray);

// Use as: await readFileAsArrayPromise(...)
```

Furthermore, some of Node's built-in modules support both the callback style and the promise style for their API. The `node:fs` module is an example. It has a promise-based version of its entire API. Here's a version of `readFileAsArray` written using the `node:fs` promise-based API:

```
import { readFile } from 'node:fs/promises';

const readFileAsArray = async (file) => {
  const data = await readFile(file, 'utf8');
  const lines = data.trim().split('\n');
  return lines;
};
```

> You can use the `await` keyword at the top level of a Node ES module (not within an `async` function). This is useful for a few cases like dynamic imports and resource initializing. However, remember that the `await` keyword pauses the execution of its scope, so using it at the top level will block all the code that comes after it.

Another way to implement handler functions is using event emitter objects, but before we talk about that, let me give a quick analogy about callbacks and promises, and about the control and trust problem I mentioned earlier.

An Analogy for Promises

When you're in a store and you order something that needs time to be prepared, they take your order and name (and money) and tell you to wait to be called when your order is ready. After a while, they call your name and give you what you ordered. The name you give them is like the callback function that you give to an asynchronous operation. It gets called when the object that was requested is ready.

The callback pattern has many disadvantages. The promise pattern is better. The superiority of promises over callbacks is all about trust and control. Let me explain that with another analogy.

Let's say you're following a recipe to make mansaf, a delicious Levantine dish of rice, lamb, and cooked yogurt. Cooking yogurt requires continuous stirring; if you stop stirring, it burns. This means that while you're stirring the yogurt, you can't do anything else (unless you turn the stove off and cancel the process).

When the yogurt starts boiling, the recipe at that point calls for lowering the heat, adding meat broth, and stirring some more.

If you're the only one cooking, you'll need to do the yogurt-stirring task synchronously. Your body, which is comparable to the single JavaScript thread in this analogy, is blocked for the duration of this synchronous task. You'll have to finish the yogurt cooking before you can start on the rice. This is similar to having synchronous loops in code:

```
putYogurtInPot();
putPotOnStove();

while (yogurtIsNotBoiling) {
  stirYogurt();
  // Now you're blocked until the loop is done
}

lowerHeat();
addMeatBroth(yogurt)

while (yogurtIsNotBoiling) {
  stirYogurt();
  // You're blocked again until the loop is done
}

// Start on the rice
```

If you need to cook both the yogurt and rice simultaneously, you need to get some help. You need to delegate! You need a sous-chef.

Having someone else do the stirring here is like having a module (like node:fs) do some async work for you (like reading the content of a file). Just like you use a callback function with an asynchronous call in Node, you need to use a callback function for your sous-chef to give them instructions about their specific tasks (and any ingredients they need):

```
sousChef.handleStirring(rawYogurt, (problem?, boilingYogurt) => {
  lowerHeat();
  const yogurtMeatMix = addMeatBroth(boilingYogurt);
  handleStirring(yogurtMeatMix) // Another async operation
});

// Start on the rice
```

By doing that, you free your single-threaded body to do something else. You can cook the rice now. You are using an asynchronous API.

So what exactly is the problem here? It's the fact that you lose control of what happens to the yogurt. Not only is the stirring process itself now controlled by your sous-chef, but the tasks that need to be done when the yogurt gets to a boiling point are also controlled by them. There is no guarantee that they will actually perform your instructions exactly like you described them.

Do you trust that they'll correctly identify the boiling point? Will they remember to add meat broth? Will they add enough and not overdo it? Will they remember to lower the heat?

To improve on this process, you can have the sous-chef announce when they think the yogurt is boiling, and then you can confirm and watch them add the meat broth. With these changes, the process becomes a promise-based one, as you basically made your sous-chef promise to announce when the boiling yogurt is ready:

```
try {
  const boilingYogurt = await sousChef.handleStirringP(rawYogurt);
  sousChef.lowerHeat();
  const yogurtMeatMix = sousChef.addMeatBroth(boilingYogurt);
  const cookedYogurt = await sousChef.handleStirringP(yogurtMeatMix);
} catch(problem) {
  sousChef.reportIt();
}

// Start on the rice
```

The only difference between handleStirring and handleStirringP is that you were promised an outcome for handleStirringP, and you can do something with that outcome. You have a little bit of trust and control added to the process. You can add the meat broth yourself (as the blocking part was really only the stirring process):

```
try {
  const boilingYogurt = await sousChef.handleStirringP(rawYogurt);
  you.lowerHeat();
  const yogurtMeatMix = you.addMeatBroth(boilingYogurt);
  const cookedYogurt = await sousChef.handleStirringP(yogurtMeatMix);
} catch(problem) {
  you: inspect(problem) && maybe(retry) || orderSomeTakeout();
}

// Start on the rice
```

More trust does not mean less control or vice versa. This really depends on the nature of the async module you use and its API.

To see the difference in syntax between nesting, chaining, and using async/await, let's cook the rice! One way to do so is to soak dry rice in water for a while. You then put the wet rice on the stove, cover it with water, and bring the water to a boil. You'll then need to lower the heat, cover the lid, and steam the rice for a while.

All of these operations are asynchronous. While you're waiting on the rice to soak, you can do something else, and so on. These operations also depend on each other. You can't do them in parallel. You still need to do them in sequence.

Because you start with a dry-rice object and that same object becomes wet rice (and then boiled rice, and then steamed rice), you can compare this process to working with promises:

```
// Nesting
dryRice.soak()
  .then(wetRice =>
```

```
    wetRice.boil()
      .then(boiledRice =>
        boiledRice.steam()
          .then(steamedRice => serve(steamedRice));
      );
  );

// Chaining
dryRice.soak()
  .then(wetRice => wetRice.boil())
  .then(boiledRice => boiledRice.steam())
  .then(steamedRice => serve(steamedRice));

// async/await
const wetRice = await dryRice.soak();
const boiledRice = await wetRice.boil();
const steamedRice = await boiledRice.steam();

serve(steamedRice);
```

The Event Loop

How exactly does a handler function get executed in the main thread after its asynchronous operation is done outside the main thread?

More importantly, when there are multiple asynchronous operations and multiple handler functions, how are they managed, and in what order do they get back to the main thread?

To answer these questions, we need to understand three main structures: the *call stack*, *event queues*, and the *event loop*.

V8 manages the execution of JavaScript functions using a simple stack structure known as the call stack. A stack is a *last-in, first-out* (LIFO) data structure where the last entity pushed to it is the first entity to be processed.

Every time a function is called, a reference to it gets placed on top of the call stack. When a stacked function calls other functions (including themselves recursively), a reference to each nested function gets placed on top of the call stack.

When a function call is completed, its reference gets popped out of the call stack. Nested function calls get popped out one at a time (from the top of the call stack) to complete the initial call of the first function that was stacked.

Any JavaScript code you write in Node has to be placed in the call stack for V8 to execute it. The call stack is single-threaded, which means when there are functions in the call stack, everything else (including handler functions for asynchronous operations) will have to wait until the call stack is available again. That's why it's important to never write code that will keep the call stack busy (like a long-running for loop).

As long as the call stack is busy, all your asynchronous operation handlers will be waiting. Any code that needs to run for a long time should be executed outside of the main thread and its one call stack.

While stacking functions in the call stack, when Node gets to an asynchronous task, it bypasses the call stack and processes the task internally. Depending on the nature of the task, Node might use a different thread or utilize the underlying OS asynchronous features.

> Node uses the asynchronous features of its underlying OS to perform non-blocking I/O operations. For CPU-bound tasks and other tasks that can't be handled asynchronously by the OS, Node provides a thread pool via its libuv library. This library is the backbone of everything asynchronous in Node. It's where the event loop is implemented, and it provides abstractions of platform-specific details so that Node can run consistently on different operating systems.

When an asynchronous task is completed, Node places its associated handler function in a queue structure known as the event queue (or task queue). A queue is a *first-in, first-out* (FIFO) structure where the first thing added is the first one processed.

For a handler function to run, it needs to be placed on the call stack. That's the job of the event loop (see Figure 3-3). It's a simple infinite loop with a simple job: when the event queue has handler functions and the call stack is empty, it picks the first queued function in the event queue and puts it on the call stack to have V8 execute it. Then it waits until the call stack is empty again to process the next queued handler function and keeps repeating that until there are no more handler functions in the event queue to process.

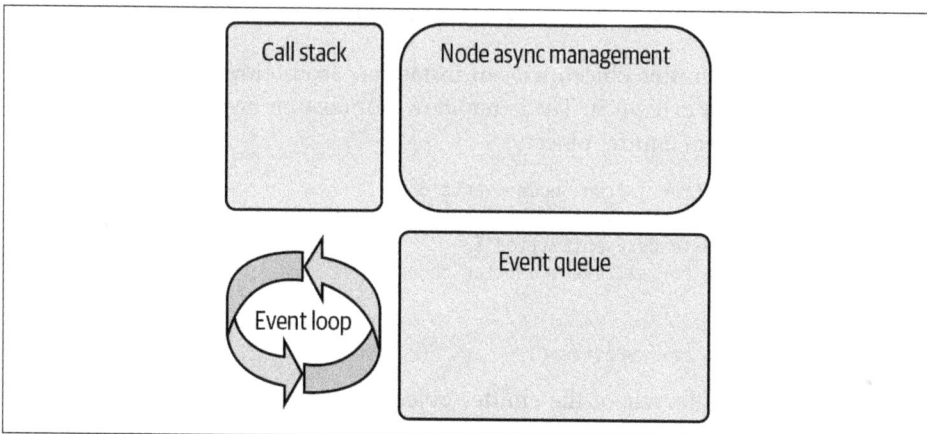

Figure 3-3. Async management and the event loop

This flow happens for any asynchronous operation, whether it's handling an incoming request to a web server, reading a file from the filesystem, starting a timer, or encrypting data. Every asynchronous task has a handler function that gets queued in an event queue, picked up by the event loop, pushed on to the call stack, then executed by V8.

> The event loop has many phases with different priorities, and it works with multiple event queues associated with these phases. There is a phase for timers, another for system-level callbacks, one for I/O operations, and a few more. Promises and high priority tasks are handled in their own queue known as the *microtask queue*.

Another option to organize and use asynchronous code in Node is through the use of event emitter objects and event listener functions. Let's talk about that next.

Event Emitters

Using events to handle asynchronous tasks takes the game to a completely different level. Events are all about good communication. No matter how complex a situation is, good communication makes things a lot better. This is true in life, and it's true with Node events.

The concept in Node is simple: emitter objects emit named events that cause previously registered handler functions to be called.

An emitter object basically has two main features:

- Emitting named events
- Registering (and unregistering) listener functions

To create an event emitter object, we can instantiate from either the `EventEmitter` class or any class that extends it. The latter allows for custom logic that can be shared between multiple event emitter objects:

```
import { EventEmitter } from 'node:events';

class MyEmitter extends EventEmitter {
  // Custom logic for all event emitter objects
}

const myEmitter = new MyEmitter();
```

At any point in the lifecycle of the emitter objects, we can use the `emit` function to emit any named event we want:

```
myEmitter.emit('something-happened', data);

// Or within the MyEmitter class:
this.emit('something-happened', data);
```

We can optionally include data objects after the event-name argument. You can use as many more arguments there as needed, and all of them will be included with the emitted event and made available to the registered handler functions.

We can add handler functions using the on method, and they will be executed every time the emitter object emits their associated name event:

```
myEmitter.on('something-happened', (data) => {
  // Do something with data
})
```

To see events in action, here's the `readFileAsArray` example implemented using events instead of callbacks or promises:

```
import fs from 'node:fs';
import { EventEmitter } from 'node:events';

class ReaderEmitter extends EventEmitter {
  readFileAsArray(file) {
    fs.readFile(file, (err, data) => {
      if (err) {
        this.emit('error', err);
        return;
      }
      const lines = data.toString().trim().split('\n');
      this.emit('data', lines);
    });
  }
}

const reader = new ReaderEmitter();

reader.on('data', (lines) => {
  const numbers = lines.map(Number);
  const oddNumbers = numbers.filter((n) => n % 2 === 1);
  console.log('Odd numbers count: ', oddNumbers.length);
});

reader.on('error', (err) => {
  throw err;
});

reader.readFileAsArray('./numbers.txt');
```

While this is a simple example, you can still see how processing the data gets its own function, and so does the error handling. This is more modular and easier to manage. As the code gets more complex, the event-based approach improves the reusability and scalability of the code. We can create multiple emitters, add more events, and handle each of the events multiple times. In addition to that, the event-based approach simplifies writing focused unit tests for the code, and it makes debugging the code easier as well.

Asynchrony

Just like with the callbacks style, having an emitter object does not mean that events are triggered asynchronously. Let's take a look at an example:

```
import { EventEmitter } from 'node:events';

class WithLog extends EventEmitter {
  execute(taskFunc) {
    this.emit('begin');
    taskFunc();
    this.emit('end');
  }
}

const withLog = new WithLog();

withLog.on('begin', () => console.log('About to execute'));
withLog.on('end', () => console.log('Done with execute'));

withLog.execute(() => console.log('*** Executing task ***'));
```

The class `WithLog` defines one instance function, `execute`. This `execute` function receives one argument, a task function, and emits events before and after its execution.

To see the sequence of what will happen here, we define listeners for both named events and execute a sample task to trigger the events. Here's the output of that:

```
About to execute
*** Executing task ***
Done with execute
```

What I want you to notice about this flow is that it all happens synchronously. There is nothing asynchronous about this code, and because this code is designed as a synchronous set of lines, if we pass in an asynchronous `taskFunc` to `execute`, the events emitted will no longer be accurate. We can simulate the case with a `setImmediate` call:

```
// ...

withLog.execute(() => {
  setImmediate(() => {
    console.log('*** Executing task ***')
  });
});
```

The output we get for that will not be correct:

```
About to execute
Done with execute
*** Executing task ***
```

To emit an event after an asynchronous function is done, we'll need to combine callbacks (or promises) with the event-based communication. Here's an example:

```
import fs from 'node:fs';
import { EventEmitter } from 'node:events';

class WithTime extends EventEmitter {
  execute(asyncFunc, ...args) {
    this.emit('begin');
    console.time('execute');
    asyncFunc(...args, (err, data) => {
      if (err) {
        return this.emit('error', err);
      }

      this.emit('data', data);
      console.timeEnd('execute');
      this.emit('end');
    });
  }
}

const withTime = new WithTime();

withTime.on('begin', () => console.log('About to execute'));
withTime.on('end', () => console.log('Done with execute'));

withTime.execute(fs.readFile, import.meta.filename);
```

The WithTime class executes an asyncFunc and reports the time that's taken by that asyncFunc using console.time and console.timeEnd calls. It emits the right sequence of events before and after the execution.

We test a withTime emitter by passing it an fs.readFile call, which is an asynchronous function. Instead of handling file data with a callback, we can now listen to the data event.

When we execute this code, we get the right sequence of events, as expected, and we get a reported time for the execution:

```
About to execute
execute: 4.507ms
Done with execute
```

Note how we needed to combine a callback with an event emitter to accomplish that. If the asynFunc supported promises as well, we could use the async/await feature to do the same:

```
class WithTime extends EventEmitter {
  async execute(asyncFunc, ...args) {
    this.emit('begin');
    try {
      console.time('execute');
      const data = await asyncFunc(...args);
      this.emit('data', data);
      console.timeEnd('execute');
      this.emit('end');
    } catch(err) {
      this.emit('error', err);
    }
  }
}
```

Errors

The error event is a special one. It should always be handled. Since we did not define a handler function for the WithTime error event, let's see what happens if we call the execute method with a bad argument. Add the bad call before the good one:

```
class WithTime extends EventEmitter {
  // ...
}

// ...

withTime.execute(fs.readFile, ''); // bad call

withTime.execute(fs.readFile, import.meta.filename);
```

The first .execute call will trigger an error. The Node process will crash and exit:

```
events.js:163
      throw er; // Unhandled 'error' event
      ^
Error: ENOENT: no such file or directory, open ''
```

The second .execute call will be affected by this crash and will potentially not get executed at all.

If we register a listener for the special error event, the behavior of the Node process will change. Let's do that:

```
// ...

withTime.on('error', (err) => {
  // Do something with err, for example log it somewhere
```

```
      console.log(err)
    });

    withTime.execute(fs.readFile, ''); // BAD CALL

    withTime.execute(fs.readFile, import.meta.filename);
```

With the special `error` event handled, the first `.execute` call will not cause the Node process to exit. The other `.execute` call will finish normally:

```
{ Error: ENOENT: no such file or directory, open '' errno: -2, code:
'ENOENT' ... }
execute: 4.276ms
```

However, there is a reason the default behavior for unhandled errors is to make the Node process exit. You need to be careful when you define handlers for errors. It's OK to handle an expected error and make an exception for it to not cause the Node process to exit, but all other errors should make the process exit. Unknown errors lead to unknown states of the code.

Examples

Many of Node's built-in modules have objects in their APIs that are instances of the `EventEmitter` class. These objects emit certain events that we can handle in our code.

For example, the Node "Hello World" simple HTTP server example that we saw in Chapter 1 can be written using events:

```
import { createServer } from 'node:http';

const server = createServer();

server.on('request', (req, res) => {
  res.writeHead(200, { 'Content-Type': 'text/plain' });
  res.end('Hello World');
});

server.on('listening', () => {
  console.log('Server is running...');
});

server.listen(3000, '127.0.0.1');
```

This is possible because the `server` object is an instance of the `http.Server` class, which extends the `EventEmitter` class under the hood, and both `request` and `listening` are predefined events in that class.

Even the `req` and `res` objects (emitted as data for the `request` event) are event emitters!

In Chapter 6, we're going to learn about Node streams, which are also event emitter objects. Here is a simple example of how we can read the content of a file with a stream object:

```
import { createReadStream } from 'node:fs';

const readStream = createReadStream('input.txt', 'utf8');

readStream.on('data', (chunk) => {
  console.log('Read chunk:', chunk);
});

readStream.on('end', () => {
  console.log('Reading finished');
});
```

Note how with this example we're reading the file one chunk at a time, instead of all at once when we use the readFile function. This is a very powerful concept in Node that'll enable us to work with big objects in a memory-efficient way. We'll learn all about that in Chapter 6.

Other Node built-in modules that have event emitter objects include node:child _process, node:net, node:dgram, node:os, and node:cluster. We'll see examples of these throughout the book.

Another way to handle emitted error events is to define a handler for the uncaught Exception event that's defined on the process object. This event is triggered for all unhandled errors. It's generally a bad practice to use it, unless you need to do something before letting the process exist as it should:

```
process.on('uncaughtException', (err) => {
  // An error was unhandled. Process should exist.

  // Do your thing

  // But force exit the process after
  process.exit(1);
});
```

However, multiple error events might happen at the same time. This means the uncaughtException listener shown might be triggered multiple times, which is a problem as any code you write in that handler will be executed multiple times.

Event emitter objects have a once method that can be used instead of the on method. Handler functions registered with the once method will be invoked only one time even if their event is triggered multiple times. This is a better method to use with the uncaughtException event to make sure that any code you write in that handler is executed only once, as we know that the process is going to exit anyway.

If multiple handler functions are registered for the same event, the invocation of those functions will be in order. The first handler function that we register will be the first function to get invoked when the event is triggered:

```
withTime.on('data', (data) => {
  console.log(`Length: ${data.length}`);
});

withTime.on('data', (data) => {
  console.log(`Characters: ${data.toString().length}`);
});

withTime.execute(fs.readFile, import.meta.filename);
```

In this example, the `Length:` line will appear before the `Characters:` line, because that's the order in which we defined those listeners.

If you need to define a new listener but have that listener invoked first, you can use the `prependListener` method:

```
withTime.on('data', (data) => {
  console.log(`Length: ${data.length}`);
});

withTime.prependListener('data', (data) => {
  console.log(`Characters: ${data.toString().length}`);
});

withTime.execute(fs.readFile, import.meta.filename);
```

In this version, the `Characters:` line will be logged first.

Finally, if you need to remove a listener, you can use the `removeListener` method.

Summary

Node's event-driven approach is a simple yet powerful concept that allows for great flexibility and control over what to do before, during, and after asynchronous operations.

We can use callbacks, promises, or event emitters to work with asynchronous operations. Node's event loop manages asynchronous tasks using event queues.

Node has an events module that we can use to define objects that support multiple events and multiple handlers for these events. Handler functions are always listening to the events they are attached to and get executed when these events are triggered. Events are triggered to indicate a condition is met.

In the next chapter, we'll discuss how to handle errors in Node and how to debug Node applications.

CHAPTER 4

Errors and Debugging

In Node, error objects are important. They are used to communicate the existence of problems and provide context and details about them. Good handling of errors makes programs more reliable and secure.

This chapter is all about errors and how to deal with them. We'll first talk about throwing and catching errors and then discuss the different types of errors and what layered error management looks like. We'll then explore debugging Node code and discuss some preventive measures that we can use to reduce potential errors.

Throwing and Catching Errors

An error in Node is an object instantiated from the `Error` class. Sometimes you'll need to create your own error objects, like this:

```
const error = new Error('Some message');
```

This is a generic error object, which is not something you should ever use, as there is a better option that we'll discuss shortly. Node also has built-in errors, which are basically error objects already created for you. The next section covers those.

What exactly do you do with an error object? Well, regardless of the type of error and whether you created it or it was built-in, whenever you're in a place in the code where a condition should not happen, you need to signal the details of that problem to whoever is using your code. You can do that by *throwing* the error.

For example, let's say we want to create a function to calculate the square root of a number. One restriction such a function should have is to block any operation that tries to use the function with a negative number, since there is no square root for a negative number.

That case is a user input error that's expected, and if it happens, the function should provide a signal that it's being used incorrectly. One simple way to do that is to throw an error:

```
function calculateSquareRoot(number) {
  if (number < 0) {
    throw new Error(
      'Cannot calculate square root of negative number'
    );
  }

  return Math.sqrt(number);
}

calculateSquareRoot(-1);
// Error: Cannot calculate square root of negative number
```

Throwing an error will make the current operation stop completely. When you throw any errors in Node, the process will crash and exit. Here's what happens when we execute this code:

```
$ node index.js
file:///Path/To/File:3
    throw new Error(
    ^

Error: Cannot calculate square root of negative number
    at calculateSquareRoot (file:///Path/To/File:3:11)
    at file:///Path/To/File:11:1
```

This might feel wrong. Why crash an entire application because of a simple wrong usage of a single function?

The simple answer here is that it's a problem that should not be ignored and should be handled somehow. Ignoring errors might lead to unintended consequences and compromises the integrity of the application. A function that's called with the wrong input might have incorrect data propagated throughout the application. That's really not good. In some cases, it might even cause a security vulnerability. In this simple example, an unvalidated input might be the path for an exploitative attack on the application.

For errors related to the application environment, not handling them might lead to leaks and potential system resources exhaustion. It's just too much of a risk to ignore any errors. In your core modules, you should always throw errors when you're in a condition that means a problem. It's your signal that a problem should either be acknowledged and handled properly or make the application crash.

Now, the modules that use your core modules might make an exception for certain errors. For example, a module that uses your calculateSquareRoot function might decide to implement an exception for having the function called with a negative number. That can be done by *catching* the error.

Whenever you write code that uses other code (like calling a function), you need to always remember that this other code might throw errors. As the user of that code, you can decide what to do with these errors. You do that by catching the error and handling it somehow.

For example, say that you decided that when `calculateSquareRoot` gets called with a negative number, it should just print a warning, not exit the whole application. Here's how you do that:

```
try {
  let result = calculateSquareRoot(-1);
} catch (error) {
  console.error(error.message);
}
```

Because the call is wrapped in a `try...catch`, and the catch block just outputs the error to the console, using this code and calling `calculateSquareRoot` with a negative number will not crash the process. This error is now a handled exception.

The problem is, the handled exception here is not just the error for a negative number argument. It's any error within the `calculateSquareRoot` function. If the function throws any other error, they will all be caught by the `try...catch` and ignored on this level as well. We should make exceptions for only the errors that we know of. Any unknown errors should still be thrown.

To do that, we need a way to conditionally handle the error. We need to check the *type* of an error. Let's talk about error types next.

Types of Errors

Different types of errors usually have different handling strategies. Knowing these types and their underlying reasons for existence helps identify what's going on when you see them.

The most important types of errors in Node are *standard errors, system errors,* and *custom errors.* Let's explore them with some examples.

Standard Errors

Standard errors are the built-in errors that are provided by JavaScript itself. These errors are thrown by the JavaScript engine when it encounters any unexpected condition that prevents the normal flow of a program. For example, when wrong syntax is used, or a variable is referenced before it was declared, or a function is used in an invalid way.

The main standard error classes in JavaScript are SyntaxError, ReferenceError, RangeError, and TypeError. All of these are subclasses of the Error class.

SyntaxError

This is the type of error thrown when we try to run code that's not valid JavaScript. We get them if we use a reserved keyword, declare a variable with an invalid name, misplace a comma, bracket, or parenthesis, or do anything else that does not follow the syntactic rules of the language:

```
> console.log('Hello world';
console.log('Hello world';
           ^^^^^^^^^^^^^

// Uncaught SyntaxError: missing ) after argument list
>
```

ReferenceError

This is the type of error thrown when you try to use a variable that has not been declared or is not in the current scope:

```
console.log(x);
// Uncaught ReferenceError: x is not defined

let x = 5;
```

RangeError

This is the type of error thrown when you try to use a value that's not within the allowed range of values enforced by either an implementation limit or a memory limit—for example, if you use a method with an argument that's not within the accepted range:

```
(123.456).toFixed(101);
    // Uncaught RangeError: toFixed() digits argument
    //                      must be between 0 and 100
```

Calling a function recursively without an exit condition could also produce this error:

```
function f() {
  f();
}

f(); // Uncaught RangeError: Maximum call stack size exceeded
```

TypeError

This is the type of error thrown when you try to use a value of an unexpected type anywhere a type is expected. For example, the method `thing.toUpperCase()` expects `thing` to be a string, but if you try to use that method with a number, you get a `Type Error`:

```
let thing = 42;
console.log(thing.toUpperCase());

// Uncaught TypeError: thing.toUpperCase is not a function
```

Other examples of `TypeError` include trying to modify a `const` variable, accessing properties on the `null` and `undefined` objects, and using the `new` keyword with non-constructor functions, among a few more cases.

> These are the most common standard errors. There are a few more like `URIError`, which is related to handling URIs, and `EvalError`, which exists only for compatibility and is no longer thrown by modern JavaScript.

System Errors

System errors in Node are thrown when something unexpected happens on the system level. They are basically the problems that come up because of the environment and OS where a Node application is run—for example, if you try to read or write to a file that does not exist, use a network port that's already in use, send data over a closed socket, and so on.

These errors have specific codes that you can look up to understand why they happen. Here are a few common system error codes:

ENOENT
: Thrown when accessing a file or directory that does not exist.

EACCES
: Thrown when an operation does not have the right permissions. For example, if we try to write to a read-only file.

EADDRINUSE
: Thrown when an HTTP (or TCP) server fails to start because the address it's trying to use is already in use.

ECONNREFUSED
: Thrown when a connection is made to a server that's not running.

Other common system errors include ETIMEDOUT (operation timed out), ECONNRESET (connection reset by peer), ENOTFOUND (entity not found), EPERM (operation not permitted), and many more. If the error code looks like ESOMETHING, it's usually a system error code. You can see the full list of system errors using os.constants.errno, as shown in Figure 4-1.

```
TERMINAL                                                    node  + v  ⬚  🗑  ...

○  $ node
   >
   > os.constants.errno
   [Object: null prototype] {
     E2BIG: 7,
     EACCES: 13,
     EADDRINUSE: 48,
     EADDRNOTAVAIL: 49,
     EAFNOSUPPORT: 47,
     EAGAIN: 35,
     EALREADY: 37,
     EBADF: 9,
     EBADMSG: 94,
     EBUSY: 16,
     ECANCELED: 89,
     ECHILD: 10,
     ECONNABORTED: 53,
     ECONNREFUSED: 61,
     ECONNRESET: 54,
     EDEADLK: 11,
     EDESTADDRREQ: 39,
     EDOM: 33
```

Figure 4-1. System errors

These system error codes are POSIX codes. POSIX (Portable Operating System Interface) is a set of standards (based on the Unix OS) that are used for maintaining compatibility between different OSs.

System errors are part of the more general category of *operational errors*. Operational errors are the nonfatal errors that can be detected, expected, and—in most cases—handled gracefully within the code. They include system errors and other runtime errors not related to the OS, service errors related to external services (like a database), input errors (for unexpected external input), and timeout errors (when an operation takes too long), among a few others.

The next error category (custom errors) can also be considered a subset of operational errors.

Custom Errors

Custom errors in Node are the errors created by you, the developer, to handle specific conditions or scenarios in your code. The error we threw for the square root of a negative number is a generic custom error.

> Custom errors are also known as *user-specified errors* (*user* in this case refers to the developer, not the end user of the application).

You'll be working with custom errors more often than other types. We create them for specific error handling and to add additional context or properties to the errors relevant to our application's logic. For example, a custom `UserNotFoundError` might be used when something tries to access the record of a user not in the system, or a `TransactionFailedError` might be thrown when a transaction fails to complete successfully. It's not uncommon for a Node application to have tens, if not hundreds, of custom errors.

While a generic instance of the `Error` class is a custom error, you should give your custom errors their own classifications. We can do that by extending the `Error` class, and then instantiating objects out of that extended class:

```
class ValidationError extends Error {
  constructor(message, fieldName) {
    super(message);
    this.name = 'ValidationError';
    this.fieldName = fieldName;
  }
}

// To use:
// throw new ValidationError('Some message', 'some_field');
```

This custom `ValidationError` class can then be used to throw its specific error when you are validating something. For example, if you need to make sure that a user object has a `username` field, you can throw this error if it does not:

```
function validateUser(user) {
  if (!user.username) {
    throw new ValidationError('Username is required', 'username');
  }
}
```

Note how the error is specific and helpful; when it gets thrown, we'll know exactly why it did, and we can get the helpful details that's packaged in it. Here's another example:

```
try {
  validateUser({}); // An empty object is not a valid user object
} catch (error) {
  if (error instanceof ValidationError) {
    console.error(
      `Error in field '${error.fieldName}': ${error.message}`,
    );
  } else {
    console.error(`Unexpected error: ${error.message}`);
    throw error;
  }
}
```

Keep the extending of the Error class to one level, and don't create a tree of errors. If your code is split into different structures (*domain*, *api*, *app*, etc.), you can categorize your errors in different modules under these structures.

No matter how you organize your custom errors, be as detailed as you can when naming them! Is there a condition where a connection to a database can't be established? Throw a Database ConnectionError. Is there a condition where a user record could not be found using a unique ID? Throw a CouldNotFindUserFrom IdError. Note how I used noun-based and verb-based names in these two examples. Pick your preference.

Note how in this example, after making the exception for the known and expected ValidationError, I am *rethrowing* any other errors. This is very important, so let's talk about that in the next section.

Assertion Errors

Another category of errors in Node is assertion errors. These are specific errors used primarily in testing and development environments. They are usually used when the code expects something to be true but finds it false.

Assertion errors are also commonly used for writing tests and when debugging and validating assumptions during development. If an assertion starts to fail anywhere in the code, that means there's likely a bug or a misunderstanding leading up to that failure point, which needs to be addressed before releasing any code changes.

Assertion errors in Node are thrown using the `assert` module, which provides a simple set of assertion tests like the following:

`assert(value[, message])`
Is the value true?

`assert.equal(actual, expected[, message])`
Does the actual value equal the expected value?

`assert.doesNotMatch(string, regexp[, message])`
Does the `string` value match the `regexp` pattern?

When an assertion method is called, it evaluates its specific condition. If the condition is not met, the assertion method throws an `AssertionError`, which details the nature of the failure. If the condition is met, the program continues execution as normal.

There are many other assertion methods for various conditions. See the Node documentation page for the `assert` module (*https://oreil.ly/I_t_F*). We'll see examples of how to use this module in Chapter 8.

Layered Error Management

Say you have modules A, B, and C, where module B uses module A, and module C uses module B. When thinking about error management, you need to remember this kind of layered structure.

With the understanding of error types and how to create custom error objects, we can now throw distinct errors for distinct conditions and give the upper layers that are using our core modules the chance to make exceptions for these distinct errors. However, after making these exceptions, a catch block should always end in a `throw error` call to rethrow any unknown errors:

```
try {
  // Some code that might throw multiple types of errors
} catch (error) {
  if (error instanceof KnownErrorType) {
    // Handle specific known error
  } else if (error instanceof AnotherKnownErrorType) {
    // Handle another specific known error
  } else {
    // Log and rethrow unknown errors
    console.log('Encountered an unexpected error: ', error);
    throw error;
  }
}
```

Without rethrowing the error object (which is unknown after making exceptions for the known ones), we would be silently ignoring it and making the application run in an unstable or incorrect state without any indication (to all upper layers) that something has gone wrong in this layer. That also makes debugging problems a lot harder since the context of the error would be lost.

By rethrowing the error, you propagate it to the higher layers. Not rethrowing the error is basically hiding it from higher layers.

There is another way to manage errors. Instead of throwing them, we can pass them forward to other parts of the application that use our core modules and let these parts either handle them or forward them again.

The error forwarding method is commonly used in asynchronous and event-driven programming, where an error encountered in one function is not immediately handled within that function but is instead sent along with any data to the next function in the sequence.

The error-first callback style discussed in Chapter 3 is basically one way to do error forwarding. When an error occurs within a function, the error is packaged up and passed as the first argument to the next callback. This allows the next function in line to check for the presence of an error and decide how to handle it:

```
function fetchData(callback) {
  getSomeDataFromAnAPI((err, data) => {
    if (err) {
      callback(err); // Forwarding the error to the callback
      return;
    }
    callback(null, data); // No error occurred, continue as normal
  });
}

fetchData((err, data) => {
  if (err) {
    console.log('Error fetching data: ', err);
    return;
  }
  console.log('Data received: ', data);
});
```

Promise-based functions are another way to forward errors. An error within the promise implementation is forwarded using the `reject` method:

```
function fetchData() {
  return new Promise((resolve, reject) => {
    getSomeDataFromAnAPI((err, data) => {
      if (err) {
        reject(err); // Forwarding the error through rejection
      } else {
```

```
        resolve(data); // Resolving the promise if no error
      }
    });
  });
}

fetchData()
  .then((data) => {
    console.log('Data received: ', data);
  })
  .catch((err) => {
    console.log('Error fetching data: ', err);
  });
```

However, we don't need promises or callbacks to implement error forwarding. We can do it simply by making functions return either data (success) or error (failure), or even both (partial success). This can be done either by unifying the return into an object with both an `error` property and `data` property, or by using a simpler approach of returning either a `data` object or an `error` object. The latter is a bit tricky as every function would need to return two different types. I would recommend the latter approach only if the Node project is using TypeScript, which adds static typing to JavaScript. We'll explore the basics of TypeScript for Node in Chapter 10.

Here's an example of a function unifying the return into an object:

```
function fetchData() {
  try {
    let data = getSomeData();
    return { error: null, data: data };
  } catch (error) {
    return { error: error, data: null };
  }
}

const result = fetchData();
if (result.error) {
  console.log('Error fetching data:', result.error);
} else {
  console.log(result.data);
}
```

Error forwarding can be used to maintain a clean and predictable flow of both data and errors through the application's operations. However, every part of the application needs to either handle errors or forward them, just as how, without error forwarding, every part of the application needs to either handle errors or throw them. Error management has to be a style that's enforced throughout the entire application.

Debugging in Node

When a code problem needs to be investigated, there are multiple ways to do the debugging in Node. The simplest way is to log information around the problem to understand what's going on. To start, simple `console.log` statements can be used to find out where exactly the problem is happening (in case the error stack was not helpful):

```
console.log('Starting application...');
app.init();
console.log('Application initialized successfully.');
```

If you don't see the second log message, that means `app.init()` has a problem.

Further logging can be used to print out data, verify expectations, and track how things change. It's a simple yet powerful way to quickly identify issues in a development environment.

For more featured debugging, Node has a built-in command-line debugging utility that can be started using the `inspect` argument:

```
$ node inspect script.js
```

This command starts your Node application with an inspector attached, allowing you to pause execution, step through the code, and inspect variables at runtime. You can set breakpoints directly from your code by adding the `debugger;` statement, which will cause the execution to pause whenever that point is reached.

The built-in debugger is certainly useful, but it's limited. There is a more powerful and featured option. You can use the Chrome browser DevTools to debug Node applications just as easily as debugging web applications. All you have to do is start the Node process with the `--inspect` flag:

```
$ node --inspect script.js
```

You can also use the `--inspect-brk` flag to break at the start of the debugged script.

After that, you can connect to the running Node process using Chrome's *chrome://inspect* page (see Figure 4-2). The running Node process will be listed there, and when you click it, you'll have all the power of DevTools to use on your Node code: the debugger, the performance profile, the memory inspector, and the real-time console.

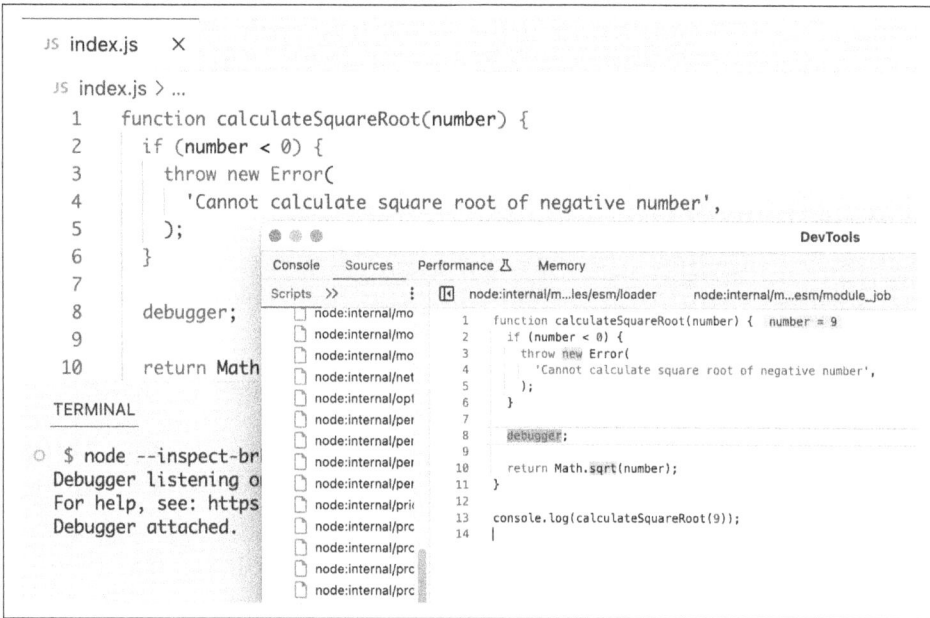

Figure 4-2. Using Chrome DevTools with Node

Node also has a built-in performance profiler that you can run with the `--prof` flag. It samples the stack at regular intervals during program execution and records the results of these samples, along with important optimization events.

Additionally, Node has multiple tracing flags that can be used to print additional information to error stacks. You can see them along with what they do in the `node` `--help` output:

```
$ node --help | grep trace
  --enable-source-maps           Source Map V3 support for stack traces
  --trace-deprecation            show stack traces on deprecations
  --trace-event-categories=...
                                 comma separated list of trace event
  --trace-event-file-pattern=...
                                 for the trace-events data, it supports
  --trace-exit                   show stack trace when an environment
  --trace-promises               show stack traces on promise
  --trace-sigint                 enable printing JavaScript stacktrace
  --trace-sync-io                show stack trace when use of sync 10 is
  --trace-tls                    prints TLS packet trace information to
  --trace-uncaught               show stack traces for the `throw`
  --trace-warnings               show stack traces on process warnings
```

Another benefit to using clear and descriptive names for objects in your code (especially functions) is that it helps with debugging. Well-named objects in stack traces and error logs make it easier to identify where and why an issue might be happening.

Preventive Measures

While dealing with errors in general is unavoidable, there are some tools and practices you can adopt to reduce the chances of having to deal with unknown errors.

Code Quality Tools

When you write a function that takes inputs, the first thing you should do is make sure that the inputs are what you expect them to be. Are they the right type? Are they in the correct format? Are they in the right range? Throw errors if they are not.

Using TypeScript with Node might at first feel like making things a lot more complicated, but in reality, the benefits you get are far greater than the little troubles.

TypeScript will point out problems in your code before the code is even run. You passed the wrong argument type to a function? You made a typo accessing a property on an object? You called a method that does not exist? TypeScript will point out all these problems, and many more, right away. TypeScript can also help you write error management code in a much better way.

The ESLint linter can also help find problems in your code before the code is run. You can use it to identify patterns that lead to bugs, and it has many other features for enforcing code style and best practices. ESLint also has a lot of powerful plugins that can extend its scope and make it work with many libraries and frameworks.

We'll expand more on both TypeScript and ESLint in Chapter 10.

Immutable Objects

A truly immutable object cannot be changed once it's created. Any updates should create new objects instead. Using immutable data structures (from a library like Immutable.js, for example) helps you avoid a ton of potential problems that come from changing data unintentionally. Even simple immutability concepts in JavaScript itself, like using `const` instead of `let`, freezing objects, and using read-only properties, are all very helpful methods to avoid many problems.

Testing

All code has to be tested one way or another. We manually test the code all the time, but we can't manually test all the cases after every change. A small change anywhere might cause bugs in places you would never expect. That's why automating tests by writing code that tests your code is really important.

Writing tests today in Node is easier than ever. Node has built-in modules for testing and assertions. The `node:test` module provides a way to organize your tests and describe them. The `node:assert` module provides assertion methods to implement the logic of the tests and throw assertion errors when the expectations are not met. Testing code in Node is the topic of Chapter 8.

Code Reviews

This goes without saying, but a solid process of reviewing all code in a project goes a long way toward finding problems early and finding better ways of doing things. I'd say every line of code should be reviewed by at least two developers.

Summary

Many problems can happen during the execution of a Node program. Some problems are out of our control, whereas others can and should be handled. Problems in Node are presented with error objects that are either built-in or customized by the developers. Core modules throw these error objects, and modules that use core modules can catch error objects and handle them if they choose to.

When an error happens, a Node process will crash and exit. This is a good thing because you don't want a program to continue running in an unknown state. However, if an error is expected, the program does not need to crash. An exception is basically an error for which we made an exception to not stop the entire program.

There are several types of errors in Node. We have standard errors from JavaScript, system errors, custom errors, and assertion errors.

In a layered application structure, errors need to propagate between the layers. That can be done using the `throw` method, or with a more structured forwarding of error objects.

Simple log messages can be used to do simple debugging in Node. Node has a built-in command-line debugging utility, and Chrome DevTools can also be used for more featured debugging of Node applications.

There are a few tools and practices that can be used to reduce the chances of having to deal with errors. Tools like TypeScript and ESLint are very powerful in that domain. Validating inputs, writing tests, using immutable data structures, and performing code reviews are all good practices to consider for that purpose too.

In the next chapter, let's talk about package management in Node and learn more about the features of Node's package manager.

CHAPTER 5

Package Management

In Chapter 1, we briefly learned about Node's default package manager, npm. It's now time to take a deeper look and get comfortable finding, using, and creating packages for Node.

The term *package* is what the software world uses to describe a folder that contains code. In Node, that folder will also have a *package.json* file that describes the metadata and dependencies of the package.

The term *module* refers to a single file or a collection of related files that encapsulate a set of functionality. Modules allow developers to organize their code into separate and reusable units. A Node package often represents a single Node module, but some packages have more than one module.

A package usually refers to *external* code that a project depends on, but I think a better word to describe package code is *generic*. You can make pieces of your own code generic and extract them into a package that you can then use in many projects.

Introducing Package Management

If packages are just folders, why exactly do we need a *package manager* for them?

Keeping track of these package folders becomes challenging when there are many of them and when these packages depend on other packages. This is especially true for a team of developers working on the same Node project. Package management tools provide a simplified and systemic approach to handling the common tasks around packages. They provide simple commands to install, update, and remove packages, and to ensure that a project has exactly what it needs to function correctly and similarly on all machines that are running it.

More importantly, package management tools can manage any conflicts among all the dependencies in the project, which are usually referred to as the *dependency tree*. It's a tree because a project has a main list of dependencies, and these main dependencies have their own dependencies, and so on. The term *transitive dependency* is often used to refer to all the dependencies in a project that are beyond the first level of the dependency tree.

> The npm tool has long been the default for managing packages and their dependencies in Node projects, but today there are a few alternatives, like Yarn, pnpm, and JSR. These alternatives to npm have their unique features and advantages. They often offer improvements on performance, disk space usage, and version management. This healthy competition has pushed npm to improve as well. In this book we're only covering npm, but you might end up using a different package manager. The basic concepts of package management are all similar, but the command interfaces and what happens behind the scenes are a bit different.

The term *npm* is mainly used to refer to the CLI (npm command) that ships with Node and provides tools to manage Node packages. There is also the npm website (*https://npmjs.com*), which hosts the public registry of many open source npm packages. The npm registry is like a big warehouse full of JavaScript packages, offering many options for common features and functionalities that you might need to add to your projects. For example, if you need your project to handle web requests, handle web sockets, or connect to a database, you don't need to build these features from scratch or deal with low-level code. You can download and use ready-made and often battle-tested generic solutions from a package registry and then build your custom needs on top of them.

But can these ready-made solutions be trusted? You need to be the judge of that, but many of these packages have already established the trust and respect of the JavaScript and Node communities. All npm packages are open source, so you can do your own research. There are bad actors in the space, so pick your packages carefully, and keep an eye on their updates. Even a trusted package might be hacked, but looking at the source code changes and the activities around the code changes (like GitHub issues, pull requests, etc.) helps mitigate the risk.

The npm registry (*https://npmjs.com*) is the default registry for the npm command, but npm is highly configurable. You can, for example, configure it to use a different registry.

Adopting a systemic approach to managing package dependencies is essential for a team project. With npm, because all package dependencies are configured in the project's *package.json* file (which is shared among all developers), it becomes easy to set up a new environment or update an older one. All developers on the team use similar versions for the project packages, and when conflicts happen, they can be detected early.

Packages usually get updated often to fix bugs, add new features, and improve things overall. With a package manager, you're in control of how to handle these updates. To ensure compatibility and prevent conflicts, you can specify which exact versions of packages the project needs. You can also automate installing important security patches.

The npm project started with a small set of Node scripts to manage common tasks around folders that contain code for Node, and it has since evolved into a fully featured package manager that is useful not only for Node but for all JavaScript code, everywhere. If you browse the packages that are hosted on the npm registry, you'll find packages that are for Node and packages that are libraries and frameworks meant to be used in a browser or a mobile application. If you dig deep enough, you'll see examples of packages for many other environments where JavaScript can be executed.

The npm registry has a lot of useless packages. Anyone can publish packages, and there is no quality control. Don't take the presence of a package on that registry as a trust signal. Always do your research, look at how the package is used in other open source projects, and preferably, inspect its code yourself.

One cool tool you can use to find actively maintained and popular packages in the Node ecosystem is the Node.js Toolbox website (*https://nodejstoolbox.com*). Top packages there are grouped by tasks, so, for example, you can see all your options for an email-delivery package or a file-upload package.

The npm Command

The npm CLI is a powerful one that supports many commands. To see usage instructions and the list of all the available commands, you can run npm --help:

```
$ npm --help

npm <command>

Usage:

npm install        install all the dependencies in your project
npm install <foo>  add the <foo> dependency to your project
npm test           run this project's tests
npm run <foo>      run the script named <foo>
npm <command> -h   quick help on <command>
npm -l             display usage info for all commands
npm help <term>    search for help on <term>
npm help npm       more involved overview

All commands:

    access, adduser, audit, bin, bugs, cache, ci, completion,
    config, dedupe, deprecate, diff, dist-tag, docs, doctor,
    edit, exec, explain, explore, find-dupes, fund, get, help,
    hook, init, install, install-ci-test, install-test, link,
    ll, login, logout, ls, org, outdated, owner, pack, ping,
    pkg, prefix, profile, prune, publish, rebuild, repo,
    restart, root, run-script, search, set, set-script,
    shrinkwrap, star, stars, start, stop, team, test, token,
    uninstall, unpublish, unstar, update, version, view, whoami
```

Don't be overwhelmed by the number of commands you see here. You don't really need many of them. The commands you'll use often are install and update. You'll also probably use run commands like start and test and a handful of other commands depending on the type of project.

Most other commands you'll use infrequently. Here are a few highlights:

npm create *(Figure 5-1)*

Sets up a new or existing npm package. This is an alias to the `init` command that we used in Chapter 1. You can use it to create a *package.json* file or execute an *initializer* for packages that provide them (like ESLint or Vite). An initializer determines how a new application is configured and built, and it can do many other tasks as well.

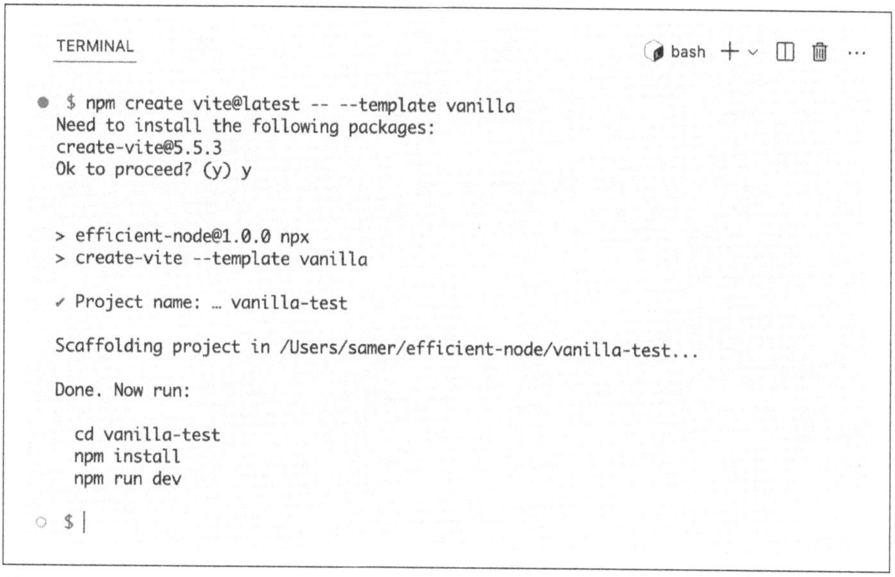

```
TERMINAL                                          bash  +  ⌄  ⬚  🗑  ⋯

 ●  $ npm create vite@latest -- --template vanilla
    Need to install the following packages:
    create-vite@5.5.3
    Ok to proceed? (y) y


    > efficient-node@1.0.0 npx
    > create-vite --template vanilla

    ✓ Project name: … vanilla-test

    Scaffolding project in /Users/samer/efficient-node/vanilla-test...

    Done. Now run:

      cd vanilla-test
      npm install
      npm run dev

 ○  $ |
```

Figure 5-1. npm create

`npm show` *<package>* (*Figure 5-2*)

Shows information about a package, such as what license it uses, how many dependencies it has, when it was published, and so on. You can use it without an argument to see information about the current package.

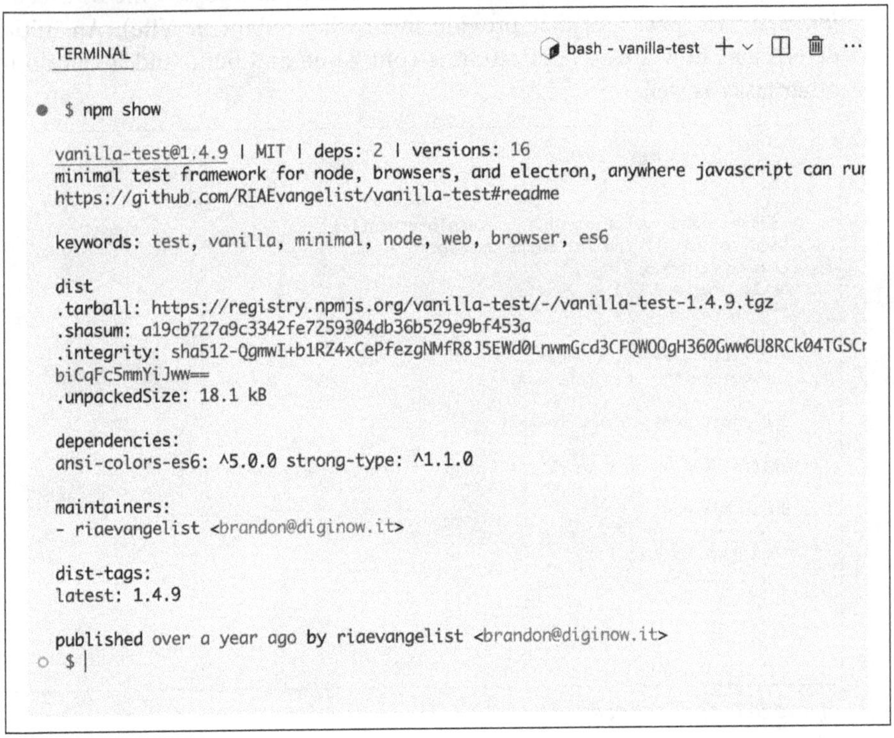

```
TERMINAL                                          bash - vanilla-test  + ∨   ▯  🗑  ⋯

●  $ npm show

   vanilla-test@1.4.9 | MIT | deps: 2 | versions: 16
   minimal test framework for node, browsers, and electron, anywhere javascript can run
   https://github.com/RIAEvangelist/vanilla-test#readme

   keywords: test, vanilla, minimal, node, web, browser, es6

   dist
   .tarball: https://registry.npmjs.org/vanilla-test/-/vanilla-test-1.4.9.tgz
   .shasum: a19cb727a9c3342fe7259304db36b529e9bf453a
   .integrity: sha512-QgmwI+b1RZ4xCePfezgNMfR8J5EWd0LnwmGcd3CFQWOOgH360Gww6U8RCk04TGSCr
   biCqFc5mmYiJww==
   .unpackedSize: 18.1 kB

   dependencies:
   ansi-colors-es6: ^5.0.0 strong-type: ^1.1.0

   maintainers:
   - riaevangelist <brandon@diginow.it>

   dist-tags:
   latest: 1.4.9

   published over a year ago by riaevangelist <brandon@diginow.it>
○  $ |
```

Figure 5-2. `npm show`

`npm search <search terms>` *(Figure 5-3)*

Searches the npm registry for packages based on the provided search query. For example, try `npm search lodash`.

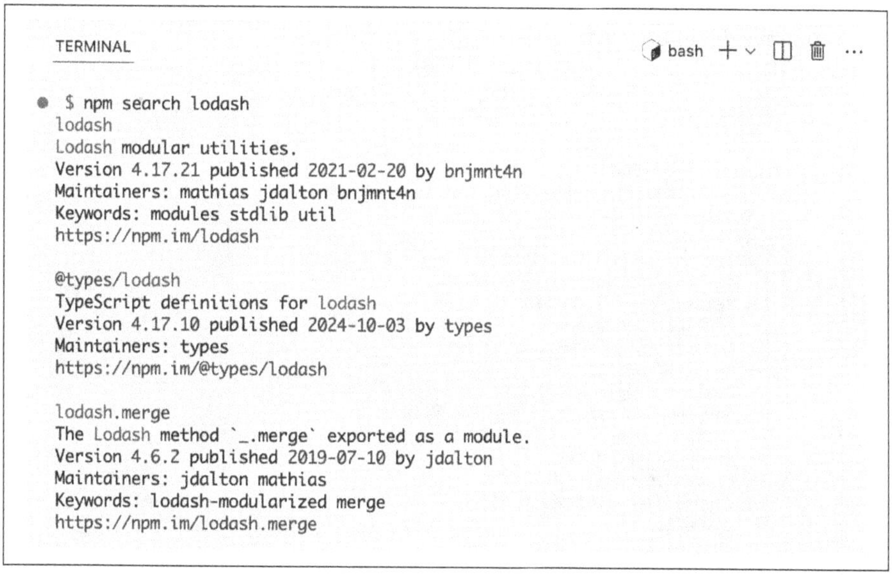

```
TERMINAL                                              bash + ∨ ⬚ 🗑 ⋯

  $ npm search lodash
  lodash
  Lodash modular utilities.
  Version 4.17.21 published 2021-02-20 by bnjmnt4n
  Maintainers: mathias jdalton bnjmnt4n
  Keywords: modules stdlib util
  https://npm.im/lodash

  @types/lodash
  TypeScript definitions for lodash
  Version 4.17.10 published 2024-10-03 by types
  Maintainers: types
  https://npm.im/@types/lodash

  lodash.merge
  The Lodash method `_.merge` exported as a module.
  Version 4.6.2 published 2019-07-10 by jdalton
  Maintainers: jdalton mathias
  Keywords: lodash-modularized merge
  https://npm.im/lodash.merge
```

Figure 5-3. npm search

npm list *<package>* *(Figure 5-4)*

Displays a tree-like view of installed packages and their dependencies, along with their versions. You can use it on a single package or for the entire application. A common alias is npm ls. Try it with a --depth=1 option to see the first level of transitive dependencies.

```
TERMINAL                                                      bash  + ⌄  ⬚  🗑  ⋯

  ●  $ npm ls --depth=1
     efficient-node@1.0.0 /Users/samer/efficient-node
     ├── @eslint/js@9.11.1
     ├─┬ eslint@9.11.1
     │ ├── @eslint-community/eslint-utils@4.4.0
     │ ├── @eslint-community/regexpp@4.11.1
     │ ├── @eslint/config-array@0.18.0
     │ ├── @eslint/core@0.6.0
     │ ├── @eslint/eslintrc@3.1.0
     │ ├── @eslint/js@9.11.1 deduped
     │ ├── @eslint/plugin-kit@0.2.0
     │ ├── @humanwhocodes/module-importer@1.0.1
     │ ├── @humanwhocodes/retry@0.3.0
     │ ├── @nodelib/fs.walk@1.2.8
     │ ├── @types/estree@1.0.6
     │ ├── @types/json-schema@7.0.15
     │ ├── ajv@6.12.6
     │ ├── chalk@4.1.2
     │ ├── cross-spawn@7.0.3
     │ ├── debug@4.3.7
     │ ├── escape-string-regexp@4.0.0
```

Figure 5-4. npm list

npm link

Creates a symbolic link between a package in your local filesystem and a package installed under *node_modules* (or globally). This allows you to develop and test packages locally without the need for publishing or reinstalling.

npm cache clean

Clears the npm cache, which can help resolve certain installation issues or outdated package versions.

npm publish

Publishes your package to the npm registry, making it available for others to install. We'll see an example of that later in the chapter.

You can get further details and instructions on any npm command using npm *<command>* -h. Here's the help summary for the npm install command:

```
$ npm install -h
Install a package
```

```
Usage:
npm install [<@scope>/]<pkg>
npm install [<@scope>/]<pkg>@<tag>
npm install [<@scope>/]<pkg>@<version>
npm install [<@scope>/]<pkg>@<version range>
npm install <alias>@npm:<name>
npm install <folder>
npm install <tarball file>
npm install <tarball url>
npm install <git:// url>
npm install <github username>/<github project>

Options:
[-S|--save|--no-save|--save-prod|--save-dev|--save-optional|..
...

aliases: add, i, in, ins, inst, ...
```

As you can see, we can use the `install` command in many ways and with many options. It also has many aliases.

You don't need to remember all of the usage ways and options, but doing a quick scan for later reference is certainly helpful.

> This is actually the summarized version of the install help page. You can see the full help page using `npm help install`.

Here are a few challenges for you to figure out from the help text of `npm install`:

- Install a package that's hosted under a scope. An npm scope is a way to group related packages under a specific namespace or organization. An example scope is `@babel`. An example package under that scope is `core`.

- Install a package directly from GitHub. Try to install `lodash` from GitHub. To verify, look at the `dependencies` section of the project's *package.json* file—`lodash` should have a `github` label.

- Install a package globally to make it available to any Node project on the machine. This option is commonly used for command-line tools. For example, you can install the ESLint package globally using npm, and that will make the `eslint` command available everywhere.

Avoid installing npm packages globally unless you really need to. Installing packages globally reduces the modularity of your projects and can lead to version conflicts between different projects. It can also cause your projects to behave inconsistently across different environments.

Semantic Versioning

The `npm update` command can be used to update packages listed in *package.json* to their latest version (as constrained in the file). To understand that, we first need to learn about *semantic versioning* (or *SemVer* for short).

SemVer is used by npm when it's time to update packages. Every package has a version. A version is one of the required pieces of information about a package, and it's usually written with the SemVer format. For example, when we installed the `lodash` package in Chapter 1, the line that was added to the *package.json* dependencies section was as follows:

```
"lodash": "^4.17.21"
```

The `4.7.21` part is the SemVer string, and it's basically a simple contract between a package author and the users of that package. When that number gets bumped up to release a new version of the package, the SemVer communicates how big of a change that new release will be to the package.

The first number, which is called the *major* number, is used to communicate that breaking changes happened in a release. Those are changes that will require users to change their code to make it work with the new release. The next time that happens for `lodash`, it'll be released with a SemVer string that begins with 5 instead of 4.

The second number, which is called the *minor* number, is used to communicate that new features were added in a release but older features should still work as is. A minor version release might also include warnings about future deprecations and API changes. Minor versions updates should still be backward-compatible and safe for users to update, without needing to make any changes to their projects.

The last number, which is called the *patch* number, is used to communicate that the release contains only bug fixes and security improvements. It should not introduce any new features or breaking changes.

You'll often see special characters before the SemVer numbers. These special characters represent a range of acceptable versions and are put to use when you instruct npm to update your dependency tree.

For example, the tilde (~) character means that an update can install the most recent patch version (remember, patch is the third number). The caret (^) character is a more relaxed constraint that means that an update can install the most recent minor version. If we update the lodash package while its version string is ^4.17.21, it'll try to find the latest version that begins with the 4 major number. So it might install a 4.19.1 package, but it will not install a 5.1.2 package.

Other special characters are =, >, >=, <, <=. If no special character is used, it means the version to be used should always be the exact one that's specified by the SemVer string. When you install an npm package, you can use the --save-exact (or -E) to instruct npm to save the dependency as an exact version without special characters.

Instead of a version string, an asterisk (*) can be used to mean the latest version available.

Another way to specify the version constraint is with an x in the string. For example, a 4.x version string means any version that begins with a 4. A 4.17.x string means any version that begins with 4.17.

You can also manually specify a range using the - character, for example: 4.15.0 - 4.17.0.

For more details on version strings and for an interactive way to test them, check out the npm SemVer calculator (*https:// semver.npmjs.com*). You can enter a version string for a particular npm package to see all the available versions constrained by that string.

I think SemVer is great. Responsible npm developers should respect it when they release new versions of their code, but it's good to treat what it communicates as a promise rather than a guarantee because even a patch release might leak breaking changes through its own dependencies. A minor version, for example, might introduce new elements that conflict with elements you previously thought were OK to use. Testing your code is the only way to provide some form of guarantee that it's not broken after an update.

Updating and Removing Packages

When the packages in your project have available updates, you can issue the npm update *<package-name>* command to update a single package, or the npm update command to update all the packages in the dependency tree.

Let's simulate a case where an update is going to happen by installing an older version of lodash. To do that, we just specify the exact version we are interested in by adding it after an @ character:

```
$ npm install lodash@3.9.1
```

You can verify which version npm installed using the `npm ls` command. It should be `3.9.1`.

Now take a look at *package.json* and note how the version string starts with a ^ character. This permits npm to update the package to the latest minor version available.

To see what version will be installed using the `npm update` command, you can first run the `npm outdated` command. It'll list all packages, and if any of them has a valid update (permitted by the version strings constraints), the updated version will be listed under the `Wanted` column. As shown in Figure 5-5, the output will also include the latest version.

```
TERMINAL                                        bash  +  ⌄  ⬚  🗑  ⋯
⊛  $ npm outdated
   Package   Current   Wanted   Latest   Location               Depended by
   lodash      3.9.1   3.10.1   4.17.21  node_modules/lodash     efficient-node
○  $
○  $ |
```

Figure 5-5. npm outdated

Now because of the ^ constraint, the `Wanted` version in this case will be `3.10.1`. That was the last `lodash` version released under the 3 major branch.

If you change the ^ into the more strict ~ and run the `npm outdated` command, the `Wanted` version will be `3.9.3`. That was the last `lodash` version released under the `3.9` minor branch.

If you change ~ into > and run the `npm outdated` command, the `Wanted` version will match the latest one.

The `outdated` command is like a dry run for you to verify what packages will be updated. It does not perform the update. To update, you run the `npm update` command.

Experiment with the `outdated`, `update`, and `ls` commands with a package like ESLint that has its own dependencies. Install an older version of that as well, for example:

```
$ npm i eslint@8
```

Note the version I use here. That 8 is the major version, and the syntax means install the latest ESLint version that begins with 8. See what version was installed with the `npm ls` command.

What happens if you change the version string in *package.json* to something older? For example, change the ESLint version string to ~7.32.0. Since that constraint specifies something older than what you currently have installed, running the npm update command will actually downgrade the ESLint package. Verify that with npm ls.

The update command will update all dependencies, including transitive ones, based on the version strings constraints specified in the *package.json* files of the packages that depend on them.

To make the outdated command show all the dependencies to be updated, run it with the -a flag:

```
$ npm outdated -a
```

Let's say that we decided we no longer want to use the ESLint package. You can remove it from *package.json* manually, but that will not remove it from *node_modules*. To remove it from both *package.json* and *node_modules*, you can run the npm uninstall <package-name>. The uninstall command is the better way here.

However, if someone on the team used the uninstall command and you pulled that code change, all you're seeing is the line being removed from *package.json*. The *node_modules* folder is not usually shared in source control repos. You'll need to run npm commands to sync your *node_modules* folder with the updates in *package.json*.

To simulate that, remove the eslint line from *package.json*. You now have packages that are installed but no longer needed (according to *package.json*). If you run the npm ls command now, it'll list these packages with an "extraneous" label next to them, as shown in Figure 5-6.

```
TERMINAL                                           bash + ∨ ⬚ 🗑 ⋯

● $ npm ls
  efficient-node@1.0.0 /Users/samer/efficient-node
  ├── @eslint-community/eslint-utils@4.4.0 extraneous
  ├── @eslint-community/regexpp@4.11.1 extraneous
  ├── @eslint/config-array@0.18.0 extraneous
  ├── @eslint/core@0.6.0 extraneous
  ├── @eslint/eslintrc@3.1.0 extraneous
  ├── @eslint/js@9.12.0
  ├── @eslint/object-schema@2.1.4 extraneous
  ├── @eslint/plugin-kit@0.2.0 extraneous
```

Figure 5-6. Extraneous packages

To remove all unused packages from the project, you can use the npm prune command:

```
$ npm prune

removed 84 packages, and audited 4 packages in 2s

1 package is looking for funding
  run `npm fund` for details

found 0 vulnerabilities
```

Now if you run the npm ls command again, there should not be any extraneous packages.

To ensure that a project's dependencies are in sync with changes in *package.json*, whenever you pull new code and notice changes to *package.json*, run both the prune and install commands.

However, the npm install command will always install the latest version of a package as permitted by version string constraint. That means in the time between a dependency being added by one developer and another developer pulling the code to install it, a new version of that dependency might have been released, and if the version string specified in *package.json* allows it, npm install will install that new version, which is different from the one that's installed on the machine that added the dependency in the first place.

That's why npm automatically maintains another file at the root of the project, the *package-lock.json* file. The purpose of that file is to lock versions of packages so that all project developers use the exact same versions of all the packages. This is true for both direct dependencies and transitive ones.

Every time a dependency is added, updated, or removed, npm will modify the *package-lock.json* file to describe the entire tree of dependencies (direct and transitive), along with what exact versions to install.

Because the *package-lock.json* file should be part of the project Git repository for others to use it, its change history can be used to go back to previous states of exactly what was under the *node_modules* folder.

The *package-lock.json* file is also used by npm to optimize its operations.

Creating and Publishing Packages

Let's create and publish a simple npm package that provides a function named `printInFrame`. That function takes a string argument and outputs that string within a frame made of * characters. Let's name the package `print-in-frame`.

Here's an example of how we'd use it:

```
import printInFrame from 'print-in-frame';

printInFrame('Hello World');
```

Running this code should output the following:

```
***************
* Hello World *
***************
```

First, make a new folder to host this package code. The name of the folder usually matches the name of the package (although that's not a requirement):

```
$ mkdir print-in-frame & cd print-in-frame
```

The next step is to make this empty folder into an npm package. We do that by adding a *package.json* file. We can use `npm init` for that:

```
$ npm init
```

Answer the questions or use the default answers. After the file is created, manually add the `'type': 'module'` to instruct Node that this project will exclusively use ES modules.

Open up your code editor and create an *index.js* file in the root of the project. To implement the `printInFrame` function, we need to read the length of the text and use that to print a set of * characters before and after. This can be done in many ways. Here's what I did:

```
// In index.js
import times from 'lodash.times';

const printInFrame = (text) => {
  const frameWidth = text.length + 4; // 2 stars + 2 spaces

  let textToPrint = '';

  times(frameWidth, () => (textToPrint = textToPrint + '*'));

  textToPrint = textToPrint + '\n' + '* ' + text + ' *' + '\n';

  times(frameWidth, () => (textToPrint = textToPrint + '*'));

  console.log(textToPrint);
```

```
    return textToPrint;
  };

  export default printInFrame;
```

I made the function depend on `lodash.times`, which provides a function that can repeat a block of code any number of times. I used that to prepare the frame header and footer lines. You need to `npm install lodash.times`.

To use the `print-in-frame` package in a Node project, we need to install it. We can actually install it directly from the filesystem. For example, if we're in a Node project that's in the same directory as the *print-in-frame* folder, in that other Node project we can do the following:

```
$ npm install ../print-in-frame
```

While this works OK, when you share your code with others, you'll have to share the *print-in-frame* folder as well. To keep them separate, we'll need to use an npm registry and publish the package there.

If you want to publish your package in the npm registry (*https://npmjs.com*), you need to have an account there. Then you can use the `npm login` command to authenticate your local npm client with your account. It'll ask you for your username and password.

Since the package name is unique at the npm registry, to avoid conflict, add a prefix to your package name. I changed the `name` property in *package.json* to `samer-print-in-frame`. While you're there, add a description to the package as well. It's optional, but it makes the package easier to discover.

When you're ready, run the `npm publish` command. If everything works, your package will be available on the npm registry website (use the UI search there to find it). You can also use the `npm search` command to find it.

With the package published, install it in your main Node project with `npm install PREFIX-print-in-frame`, replacing *PREFIX* with the prefix that you used.

Now look at the output of `npm ls`. You should see two new dependencies: *PREFIX-print-in-frame* and `lodash.times`.

To make updates to your package and test them in a project before you publish a new version, you can use the `npm link` command to temporarily make a project use a local package rather than the one installed through the registry. In the *print-in-frame* folder, run `npm link`; then in the main project folder, run `npm link PREFIX-print-in-frame`.

Now you can make changes to your local package folder and test them in your main project. Once you're done, you can increment the `version` property in the *package.json* file under your package and run `npm publish` again.

> I used ES modules for `print-in-frame`. This means it can be used only under projects that use ES modules. If you want to create a package that can be used in any Node project, you'll need to create a CommonJS version as well. You can use tools like Babel or TypeScript to automate tasks like these.

npm Run Scripts

Scripts are a feature in npm that enables developers to easily perform (or automate) common tasks like building, testing, and deploying applications.

You can define a run script under the `scripts` section in *package.json*. When you run the `npm init` command, it'll include an example run script:

```
"scripts": {
  "test": "echo \"Error: no test specified\" && exit 1"
},
```

You can use that `test` script by running `npm run test`. A few common run script names (like `test`, `start`, `stop`) have a shortcut alias as well. You can run the `test` script here with just `npm test`.

> If you run the `npm run` command without any arguments, it'll list all defined scripts under the project.

The sample `test` script just outputs an error message, but note how it used shell commands like `echo` and `exit`. You can use any of the shell commands available on your machine. For example, try a script to `ls -al` or to `npm ls | grep 'extraneous'`. The latter is a good example of how a common project task can be simplified into a run script and documented for other team members who don't know about it. What's a good intuitive name for that task? Maybe `list-unused-packages`:

```
"scripts": {
  ...
  "list-unused-packages": "npm ls | grep 'extraneous'"
},
```

Now a developer who does not know about this extraneous label can look at this run script and figure out how to list any unused packages in the project. They just need to `npm run list-unused-packages`.

This becomes more important when you publish packages for other teams to use. The best place to communicate to developers how to use a package is in npm run scripts.

Run scripts help developers automate running tasks. First, if you need to run something repeatedly for the project—for example, run all integration tests—you'll have a simple and intuitive way of doing it, rather than trying to figure out the exact command every time. More importantly, an npm run script will make running this task consistent among all developers. All developers should be using the exact same command to run all integration tests. Even more importantly, if the way to run all integration tests needs to change, instead of manually announcing this change in a chat channel, you can communicate it with a change to *package.json*, which is forever kept in the project's Git history.

You can even make the automation official and adopt a way to run tasks automatically before or after other tasks. For example, I often forget to run `npm prune && npm install` after pulling new code and trying to run all tests. You can use an npm run script to automatically run the pruning and installing every time you run the tests.

To do that, you can define script names using a `pre` or `post` prefix. For this example, we can define a `pretest` script to prune/install:

```
"scripts": {
  ...
  "pretest": "npm prune && npm install"
},
```

With that special script in place, every time you run `npm test`, the `prune/install` commands will be executed before running the tests.

This works with any script name. If you have `dosomething` script name, you can define the `predosomething` and `postdosomething` scripts to execute tasks before or after you run `dosomething`.

This is great for many use cases. You can, for example, automate running tests, formatting/linting, or generating documentations every time you try to push new code to its repo.

One other cool thing about npm run scripts is that they'll execute any command-line tools installed under the project. You don't need to explicitly specify the path to these commands.

For example, run `npm i eslint` under the project to install the `eslint` command-line. Now, if you're in the project folder and you try to execute the `eslint` command, it will not be available. That command is somewhere under the *node_modules* folder, but npm does not make it globally available. However, npm run scripts recognize the commands available under *node_modules*. To test that, add the following script:

```
"scripts": {
  ...
  "lint": "eslint"
},
```

Now you can `npm run lint` and npm will find the `eslint` command and execute it. You can even include arguments, and npm will pass them to what you're executing, but since npm commands also accept arguments, you'll need to use a `--` separator to pass arguments to the executed script. For example, to see the help page for `eslint`, this is what you need to do:

```
$ npm run lint -- --help
```

Executing the same line without the `--` arguments separator will output the help page for the `run-script` npm command. Any arguments you pass before the `--` separator will be used by the `run-script` npm command.

> I named the script `lint` (instead of `eslint`) intentionally. Generic names are better under npm scripts. Maybe in the future we'll use something other than `eslint` to lint. Changing a run script name might break things in the future, especially automated tasks.

The npx Command

Another option to execute a command-line tool that's installed in a project is `npx` (which stands for *node package execute*). For example, running `npx eslint --help` will *always* work, even if the ESLint package is not installed in the project.

If a package is installed, the `npx` command will use the local folder under *node_modules*. If the package is not installed locally, `npx` will automatically download a temporary copy of the package and use it to execute the package.

Test it with `eslint`. Uninstall it with `npm uninstall eslint`, then run `npx eslint --help` to see how `npx` will download and then run the command.

Just like `npm`, you can use `npx` with specific versions. For example, let's say that you need to find out which of the `eslint` options (which you can see in the help page) existed early on, since the first available version of `eslint`.

You can use the `npm show` command to find out the earliest available version of `eslint`:

```
$ npm show eslint versions
```

When I tested this command, the earliest version of eslint was 0.4.0. Note that an earlier version might have been available, but the maintainers of eslint decided to purge it from the registry.

To see the help page of the 0.4.0 eslint command, you can run npx eslint@0.4.0 --help, as shown in Figure 5-7.

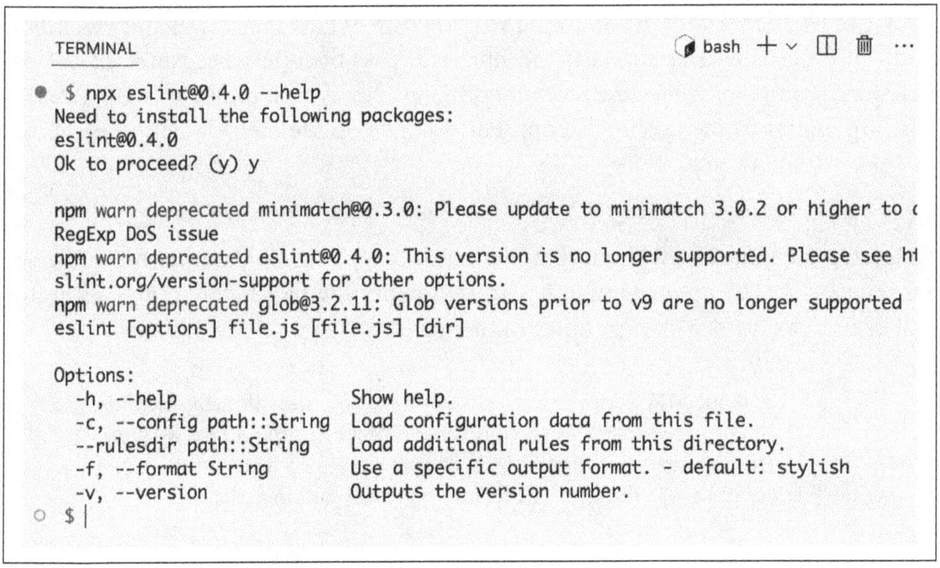

Figure 5-7. Using npx with a version

The npx command is commonly used to bootstrap a project from a template. An example of a package that can be used that way is create-react-app. You can use it through the npx command to generate a working React application using one of the many supported templates:

```
$ npx create-react-app your-app-name-here
```

Not only will this download a temporary copy of the create-react-app package, it'll then recognize that this is a *generator* package with a default command to create a project. It'll execute that default command.

Generator packages can even have multiple commands. Checkout the help page for the @vue/cli generator package to see an example of that:

```
$ npx @vue/cli --help
```

Summary

A package manager like npm is an important part of working on a Node project. It introduces a simple and standard way to deal with project external dependencies and keep them updated, consistent, and conflict-free.

With npm, packages are hosted on a public registry, and the npm command is configured to work with that registry. Related npm commands like install, update, search, and more also work with that registry.

The *package.json* and *package-lock.json* files are automatically modified by npm every time there is a change to the project dependency tree. These files store what versions of packages are installed and what range of versions should be used when updating packages.

In addition to the npm command, there's also an npx command that can be used to execute local or remote command-line tools.

In the next chapter, we'll explore how to work with streams in Node to efficiently handle large amounts of data without exhausting the process.

CHAPTER 6

Streams

Node streams have a reputation for being hard to understand and work with. That might have been the case in the early days of Node, but things have changed. Today, it's relatively easy to create and consume streams in Node. We can even use native JavaScript async iterators and generators to work with Node streams.

In this chapter, I'll explain the concept of streams, why they're needed, and how to create, use, and combine them to efficiently process large amounts of data without overwhelming the memory available to a Node process.

Introducing Streams

When you download a file from the internet, watch a show, or listen to a song, you are using streams. The content is being streamed to you one chunk at a time.

Streams are basically collections of data, similar to arrays or strings, but instead of storing data in memory space, streams process data over time. You can use streams to process very large amounts of data using a limited memory space.

Analogies for streams are all around us in life. When you have a sink full of dirty dishes, you can load them up in a dishwasher, and that would be processing all of them at once, like holding them in an array. If you can't use a dishwasher, you'll need to employ the concept of streams and clean up the sink one dish at a time. You take one dish, scrub it, rinse it, and then move on to the next one.

This is great if you have a small number of dishes to wash. If you have a lot of dishes to wash and someone who can help, you can use two streams! You can have one person scrub and the other rinse. Each person deals with a different task, but the input of one process (the rinsing) depends on the output of another (the scrubbing). We can add more streams from there, such as drying and stacking. That's how multiple streams can be combined to make tasks more efficient.

When we use the output of one stream as the input for another, that's known as *piping*. That's the same term used in the Unix OS as *pipes* are used to transfer data from one process to another.

Here's an example for using pipes in Linux. Suppose you are tasked with counting the occurrence of the word `require` in all files under a big project. In Linux, you have the `grep` command that can find patterns in files and the `wc` command that can count lines, words, or characters. We can use the output of the `grep` command and pipe it as the input of the `wc` command:

```
$ grep -wR require * | wc -l
```

Pipes in Linux enable us to compose powerful commands from smaller commands, and the ability to pipe streams on each other gives them the power of composability as well.

If we had a stream to `grep` and a stream to `wc` in Node, we can combine them like this:

```
const grep = ... // A stream for the grep output
const wc = ...   // A stream for the wc input

grep.pipe(wc)
```

Or, to put the dishwashing analogy in pseudocode:

```
const scrub = ... // An output stream of scrubbed dishes
const rinse = ...   // An input stream for dishes to be rinsed

scrub.pipe(rinse)
```

We'll talk more about the `pipe` method later in the chapter.

Before we conclude the dishwashing analogy, let's use it to understand a couple more concepts about streams.

There are two major types of streams in Node: readable streams and writable ones. Some streams are both readable and writable; these are known as duplex streams.

A readable stream produces output, and a writable stream needs input. In this analogy, the scrub process is the readable part (it has an output of scrubbed dishes), and the rinse process is the writable part (it needs an input of scrubbed dishes). When using the pipe method in Node, the caller has to be a readable stream, and the argument has to be a writable stream:

```
readableStream.pipe(writableStream)
```

Data flow in streams is another concept you should understand. As long as the two people washing the dishes are handling their tasks at a relatively similar rate, the flow of dishes from dirty to scrubbed to rinsed works fine. But let's say the scrubber needs to work on one particularly dirty dish and scrub it really well; the rinser in that case might need to pause a bit before they get the next dish. Similarly, if the rinser is behind, a pile of scrubbed dishes might start to accumulate, and the scrubber might need to pause. These are some of the challenging aspects of using streams. We'll talk more about that shortly.

Using Streams

To really understand the powerful efficiency of streams, let's see some numbers. Let's go over one practical example of doing the same task with and without streams, and compare memory usage.

First, we need a big file. Streams are all about processing large amounts of data. Here's a simple way to create a big file in Node:

```
import fs from 'node:fs';
import { randomBytes } from 'node:crypto';

const file = fs.createWriteStream('./big.file');

for (let i = 0; i <= 1e6; i++) {
  file.write(
    randomBytes(200).toString('hex')
  );
}

file.end();
```

Note what I used to create that big file: a writable stream! The node:fs module can be used to read from and write to files using a stream interface. We created this big file one chunk at a time, using a loop. In every iteration, we *streamed* one line to the file, and we repeated that 1 million times.

When I ran this code, it generated a 381 MB file.

Now, let's say that we need to make this file available for download from a Node web server. Here's an example web server designed to exclusively serve *big.file*:

```
import { createServer } from 'node:http';
import { readFile } from 'node:fs/promises';

const server = createServer();

server.on('request', async (req, res) => {
  const data = await readFile('./big.file');

  res.end(data);
});

server.listen(3000, () => {
  console.log('Server is running...');
});
```

When this web server gets a request, it'll serve the big file content using the `fs.read File` method. I didn't need to use streams to make that happen, and the code uses asynchronous methods, so we're not blocking the call stack. So what exactly is the problem?

Well, let's see what happens when we run the server, connect to it, and monitor the memory while doing so.

When I ran the server, it started out with a normal amount of memory for a process —17.8 MB, as shown in Figure 6-1.

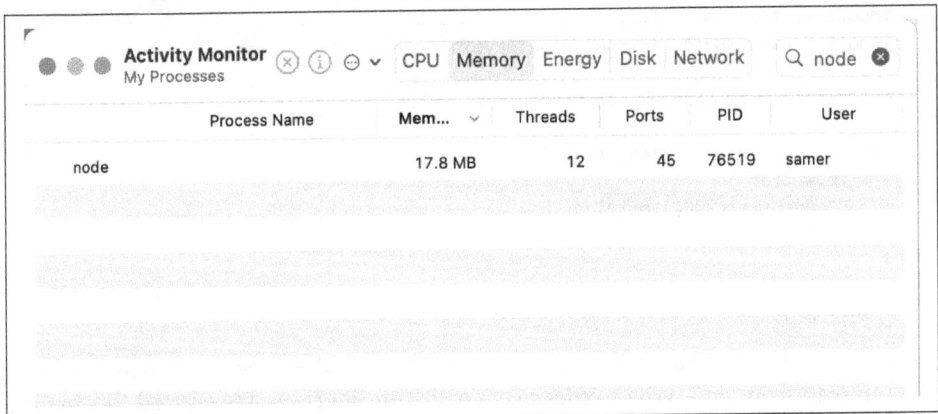

Figure 6-1. The server process's initial memory

Then I connected to the server. Figure 6-2 shows what happened to the memory consumed—it jumped to 402.4 MB.

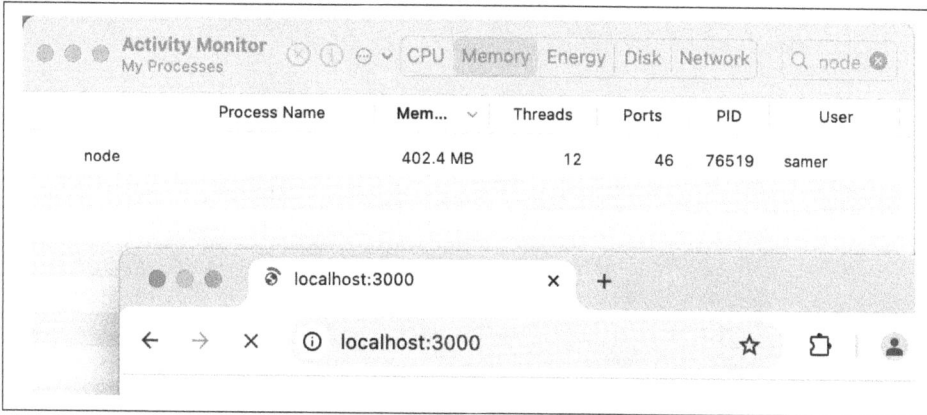

Figure 6-2. The server process's memory after a request

We basically put the whole *big.file* content in memory before we wrote it out to the response object. That is very inefficient, and we can do much better by using streams.

Luckily, many of Node's built-in modules use streams under the hood, and the `node:http` module is one of them. The `res` object in the `request` event handler is a writable stream! We can use it to stream data to the client instead of sending it all at once. We just need a readable stream to represent the content of *big.file*, and we can then use the `pipe` method to send the file one chunk at a time without consuming around 400 MB of memory.

The `node:fs` module can give us a readable stream for any file using the `createRead Stream` method. Here's how we can use it with the `res` stream object:

```
import { createServer } from 'node:http';
import { createReadStream } from 'node:fs';

const server = createServer();

server.on('request', (req, res) => {
  const src = createReadStream('./big.file');
  src.pipe(res);
});

server.listen(3000, () => {
  console.log('Server is running...');
});
```

To test that, after you run the server, connect to it to download the file, and watch the memory consumption of the process while you do (see Figure 6-3).

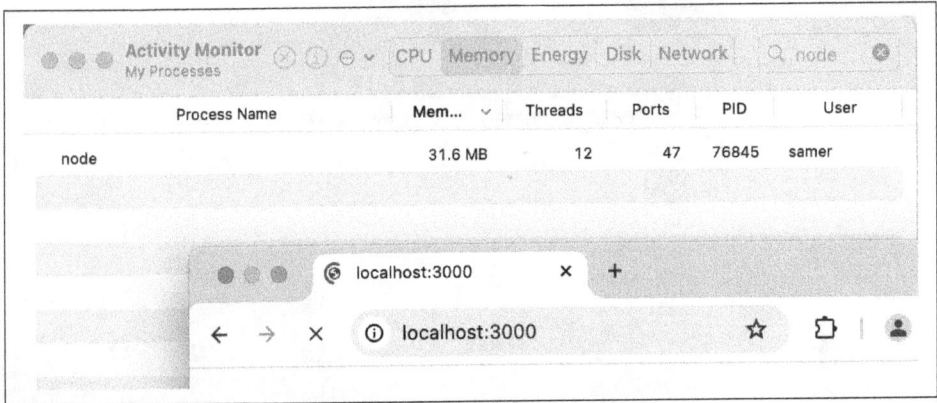

Figure 6-3. The server process's memory with streams

With this version of the web server, when a client asks for that big file, we stream it one chunk at a time, which means we don't buffer it in memory at all.

You can push this example to its limits. Regenerate *big.file* with 5 million lines instead of just 1 million, which would take the file to well over 2 GB. In most Node installations, that would be even bigger than the default buffer limit.

If you try to serve that bigger file using `fs.readFile`, you simply can't by default (you can change the limits). But with `fs.createReadStream`, there is no problem at all streaming 2 GB of data to the requester and, best of all, the process memory usage will be roughly the same.

Fundamentals of Streams

As we learned in the previous section, streams can be readable, writable, or both. A readable stream is an abstraction for a source from which data can be consumed. An example of that is the `fs.createReadStream` method. A writable stream is an abstraction for a destination to which data can be written. An example of that is the `fs.createWriteStream` method.

A stream that is both readable and writable is known as a *duplex* or *bidirectional* stream. These streams are useful where data needs to flow in both directions, such as in network communication using TCP sockets.

A transform stream is a duplex stream that can be used to modify the data in transit. An example of that is the `zlib.createGzip` stream that compresses data using gzip. You can think of a transform stream as a function where the input is the writable stream part and the output is the readable stream part.

Many of the built-in modules in Node implement the streaming interface. Table 6-1 lists some examples.

Table 6-1. Examples of stream objects

Readable streams	Writable streams
HTTP requests	HTTP responses
`fs` read streams	`fs` write streams
`zlib` streams	`zlib` streams
`crypto` streams	`crypto` streams
TCP sockets	TCP sockets
child-process `stdout` & `stderr`	child-process `stdin`
`process.stdin`	`process.stdout`, `process.stderr`

Notice that the stream objects in Table 6-1 are closely related. Also note how the `stdio` streams (`stdin`, `stdout`, `stderr`) have the inverse stream types when it comes to child processes. This allows for a really easy way to pipe to, and from, these child-process `stdio` streams using the main-process `stdio` streams. We'll learn about child processes in Chapter 7.

All streams are instances of `EventEmitter`. They emit events that can be used to read and write data. However, we can consume streams data in a simpler way using the `pipe` method.

The pipeline Method

The `pipe` method in Node connects two stream objects: a readable source and a writable destination. This is the important line that you need to remember:

```
readableSrc.pipe(writableDest);
```

In this simple line, we're piping the output of a readable stream (the source of data) as the input of a writable stream (the destination). The source has to be a readable stream, and the destination has to be a writable one. Of course, they can both be duplex/transform streams as well. In fact, if we're piping into a duplex stream, we can chain pipe calls just like we do in Linux:

```
readableSrc
  .pipe(transformStream1)
  .pipe(transformStream2)
  .pipe(finalWritableDest);
```

The `pipe` method returns the destination stream. That allows for chaining if the destination stream is readable as well.

For streams a (readable), b and c (duplex), and d (writable), we can do the following:

```
a.pipe(b)
 .pipe(c)
 .pipe(d);

// Which is equivalent to:
a.pipe(b);
b.pipe(c);
c.pipe(d);

// Which, in Linux, is equivalent to:
// $ a | b | c | d
```

The `pipe` method is the easiest way to consume streams as it automatically manages the flow of data so that a writable destination is not overwhelmed by a readable source. When we use other stream methods to read/write data, we'll need to manually manage the flow of data.

Usually, when we're using the `pipe` method, we don't need to use stream events other than the `error` events, but if we need to consume the streams in more custom ways, we can combine other stream methods and events to do that.

An even better way to consume multiple streams is to use the `pipeline` method. You can use it with both the callback and promise patterns:

```
import { pipeline } from 'node:streams/promises';

await pipeline(
  readableSrc,
  transformStream1,
  transformStream2,
  finalWritableDest
);
```

Besides the more clear syntax, the `pipeline` method is better at handling stream errors and doing any needed cleanups.

You should use the `pipeline` function within a `try/catch` statement to handle any errors thrown by any streams in the pipeline.

Stream Events

We can use stream methods like `read` and `write` in combination with stream event listeners to consume streams.

Here's the simplified equivalent code of what the pipe method mainly does to read and write data:

```
// readable.pipe(writable)

readable.on('data', chunk => {
  writable.write(chunk);
});

readable.on('end', () => {
  writable.end();
});
```

Table 6-2 shows a list of the important events and methods that can be used with readable and writable streams.

Table 6-2. Important events and methods for streams

	Readable streams	Writable streams
Events	`data, end, error, close, readable`	`drain, finish, error, close, pipe, unpipe`
Methods	`pipe(), unpipe(), wrap(), destroy()`	`write(), destroy(), end()`
	`read(), unshift(), resume(), pause(), setEncoding()`	`cork(), uncork(), setDefaultEncoding()`

The events and methods in Table 6-2 are somehow related because they are usually used together.

Events and methods can be combined for custom use of streams. To consume a readable stream, we can use the `pipe`/`unpipe` methods or the `read`/`unshift`/`resume` methods. To consume a writable stream, we can make it the destination of `pipe`/`unpipe`, or just write to it with the `write` method and call the end method when we're done.

When not using `pipe`/`unpipe`, the use of stream events will be needed. The most important events on a readable stream are as follows:

The `data` *event*
    Emitted whenever the stream passes a chunk of data to the consumer

The `end` *event*
    Emitted when there is no more data to be consumed from the stream

The most important events on a writable stream are as follows:

The `finish` *event*
> Emitted when all data has been flushed to the underlying system

The `drain` *event*
> A signal that the writable stream can receive more data

The `drain` event is important. It's needed to ensure a balanced flow of data. The rate at which data is pushed by or written to a stream is known as the *pressure* of the stream. When a writable stream has a slower rate of processing data than the rate of a readable stream that's pushing data, data will start buffering in the writable stream. This condition is known as *backpressure*.

As long as a writable stream is operating within its memory limits, its `write` method will return `true`. When the memory limit is reached, the `write` method returns `false` to indicate that further attempts to write data to the stream should stop until the `drain` event is emitted.

The `error` event may be emitted by streams at any time and should always be handled, even when using the `pipe` method:

```
readable.pipe(writable);

readable.on('error', (err) => {
  // Handle potential read errors
});

writable.on('error', (err) => {
  // Handle potential write errors
});
```

The callback functions for handling `error` events receive a single `Error` object as an argument. This object can be used to handle different errors differently.

Paused and Flowing Modes

Readable streams have two main modes that affect the way we can consume them. They can be either in the *paused* mode or in the *flowing* mode. These modes are sometimes referred to as *pull* and *push* modes.

All readable streams start in the paused mode by default, but they can easily be switched to flowing and back to paused when needed. Sometimes, the switching happens automatically.

When a readable stream is in the paused mode, we can use the `read()` method to read from the stream on demand. However, for a readable stream in the flowing mode, the data is continuously flowing, and we have to listen to events to consume it.

In the flowing mode, data can actually be lost if no consumers are available to handle it. This is why when we have a readable stream in flowing mode, we need a `data` event handler. In fact, just adding a `data` event handler switches a paused stream into flowing mode, and removing the `data` event handler switches the stream back to paused mode. Some of this is done for backward compatibility with the older Node streams interface.

To manually switch between these two stream modes, we can use the `resume()` and `pause()` methods.

> When consuming readable streams using the `pipe` or `pipeline` methods, we don't have to worry about these modes because they are managed automatically.

Implementing Streams

When we talk about streams in Node, there are two main tasks:

- *Implementing* the streams
- *Consuming* the streams

So far, we've been talking only about consuming streams. Let's implement some!

Modules that implement streams are usually the ones that import the `stream` module.

Writable Streams

To implement a writable stream, we use the `Writable` constructor from the stream module:

```
import { Writable } from 'node:stream';
```

We can implement a writable stream in many ways. For example, we can extend the `Writable` class:

```
class myWritableStream extends Writable {
  // ...
}
```

However, I prefer the simpler constructor approach. We just create an object from the `Writable` constructor and pass it a number of options. The only required option is a `write` function, which exposes the chunk of data to be written:

```
import { Writable } from 'node:stream';

const outStream = new Writable({
```

```
    write(chunk, encoding, callback) {
      console.log(chunk.toString());
      callback();
    }
});
```

```
    process.stdin.pipe(outStream);
```

This write method takes three arguments:

The chunk
>   This is the streamed data. It's usually a Node buffer object but can also be a string.

The encoding *argument*
>   This is needed if chunk is a string to be encoded with an encoding different from the default utf8. You can omit it otherwise.

The callback
>   This is a function that we need to call after we're done processing the data chunk. It's what signals whether the write step was successful or not. To signal a failure, we call the callback function with an error object as its argument.

In this example, we simply console.log the chunk (as a string) and call the callback function after that without an error (to indicate success). This is a very simple echo stream. It simply echoes back anything it receives.

To consume this stream, we used it with process.stdin, which is a readable stream, so we can just pipe process.stdin into our outStream.

When we run this code, anything we type into process.stdin will be echoed back using the outStream console.log line.

The outStream object is actually similar to the built-in process.stdout object. We can simply pipe stdin into stdout and we'll get the exact same echo feature with this single line:

```
    process.stdin.pipe(process.stdout);
```

Readable Streams

To implement a readable stream, we use the Readable constructor from the stream module:

```
import { Readable } from 'node:stream';

const inStream = new Readable();
```

There is a simple way to implement readable streams. We can just directly push the data that we want the consumers to consume:

```
// ...

inStream.push('ABCDEFGHIJKLM');
inStream.push('NOPQRSTUVWXYZ');

inStream.push(null); // No more data

inStream.pipe(process.stdout);
```

When we push a null object, that means we want to signal that the stream does not have any more data.

To consume this simple readable stream, we simply piped it into the writable process.stdout stream object.

When we run this code, we'll be reading all the data from inStream and echoing it to the standard output. Very simple, but also not very efficient. We're basically pushing all the data in the stream all at once and then piping it to process.stdout. The much better way is to push data on demand, when a consumer asks for it. We can do that by implementing the read() method in a readable stream configuration:

```
const inStream = new Readable({
  read(size) {
    // There is a demand for the data...
    // Someone wants to read it.
  }
});
```

When the read method is called on a readable stream, the implementation can push partial data to the queue. For example, we can push one letter at a time, starting with character code 65 (which represents A), and increment the code on every push:

```
const inStream = new Readable({
  read(size) {
    this.push(String.fromCharCode(this.currentCharCode++));
    if (this.currentCharCode > 90) {
      this.push(null);
    }
  },
});

inStream.currentCharCode = 65;

inStream.pipe(process.stdout);
```

While the consumer is reading a readable stream, the `read` method will continue to fire, and we'll push more letters. We need to stop this cycle somewhere, and that's why I used an `if` statement to push `null` when the `currentCharCode` is greater than `90` (which represents `Z`).

This code is equivalent to the simpler one we started with, but now we're pushing data on demand when the consumer asks for it, as shown in Figure 6-4. You should always do that.

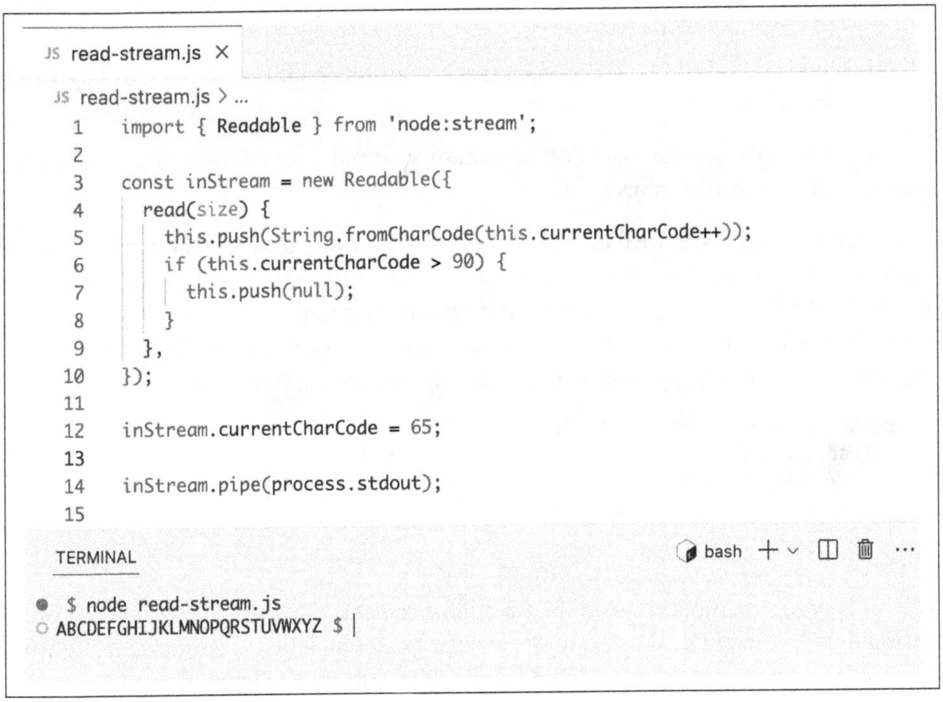

```
JS read-stream.js ✕

JS read-stream.js > ...
 1    import { Readable } from 'node:stream';
 2
 3    const inStream = new Readable({
 4      read(size) {
 5        this.push(String.fromCharCode(this.currentCharCode++));
 6        if (this.currentCharCode > 90) {
 7          this.push(null);
 8        }
 9      },
10    });
11
12    inStream.currentCharCode = 65;
13
14    inStream.pipe(process.stdout);
15
```

```
TERMINAL                                          bash + ∨ ⊓ 🗑 ⋯

● $ node read-stream.js
○ ABCDEFGHIJKLMNOPQRSTUVWXYZ $ |
```

Figure 6-4. A readable stream

If the source of data is a string, an array, or any other iterable object, you can use the `Readable.from` method to create a readable stream for it:

```
// To convert string data into a readable stream
Readable.from('ABCDEFGHIJKLMNOPQRSTUVWXYZ');
```

You can use JavaScript function generators with `Readable.from` and use the `yield` keyword to make more data readable with the created stream. Here's an example:

```
function* generate() {
  for (let index = 65; index <= 90; index++) {
    yield String.fromCharCode(index);
  }
}
```

```
// To create a stream from yielded values:
Readable.from(generate());
```

Duplex/Transform Streams

With `Duplex` streams, we can implement both readable and writable streams within the same object. It's as if we inherit from both interfaces.

Here's an example duplex stream that combines the preceding writable and readable examples:

```
import { Duplex } from 'node:stream';

const inoutStream = new Duplex({
  write(chunk, encoding, callback) {
    console.log(chunk.toString());
    callback();
  },

  read(size) {
    this.push(String.fromCharCode(this.currentCharCode++));
    if (this.currentCharCode > 90) {
      this.push(null);
    }
  }
});

inoutStream.currentCharCode = 65;

process.stdin.pipe(inoutStream).pipe(process.stdout);
```

By combining the methods, we can use this duplex stream to read the letters from A to Z, and we can also use it for its echo feature. We pipe the readable `stdin` stream into this duplex stream to use the echo feature, and we pipe the duplex stream itself into the writable `stdout` stream to see the letters A through Z.

It's important to understand that the readable and writable sides of a duplex stream operate completely independently from one another. This is merely a grouping of two features into one object.

A transform stream is a more interesting duplex stream because its output is computed from its input. For a transform stream, we don't have to implement the `read` or `write` methods; we only need to implement a `transform` method, which combines both of them. It has the signature of the `write` method, and we can use it to push data as well.

Here's a simple transform stream that echoes back anything you type into it after transforming it to uppercase format:

```
import { Transform } from 'node:stream';

const upperCaseTr = new Transform({
  transform(chunk, encoding, callback) {
    this.push(chunk.toString().toUpperCase());
    callback();
  }
});

process.stdin.pipe(upperCaseTr).pipe(process.stdout);
```

In this transform stream, which we're consuming exactly like the previous duplex stream example, we implemented only a transform() method. In that method, we convert the chunk into its uppercase version and then push that version as the readable part, as shown in Figure 6-5.

```
JS transform-stream.js ✕

JS transform-stream.js > ...
   1    import { Transform } from 'node:stream';
   2
   3    const upperCaseTr = new Transform({
   4      transform(chunk, encoding, callback) {
   5        this.push(chunk.toString().toUpperCase());
   6        callback();
   7      },
   8    });
   9
  10    process.stdin.pipe(upperCaseTr).pipe(process.stdout);
  11
```

TERMINAL   node + ∨ ⬚ 🗑 ⋯

```
○ $ node transform-stream.js
Hello World!
HELLO WORLD!
```

Figure 6-5. A transform stream

Async Generators and Iterators

We saw how a generator function can be used to create a readable stream. This works with async iterators as well, so you can use for await in an async generator function and use that function to create a readable stream.

Here's an example of a readable stream created from API calls to GitHub:

```
import { Readable } from 'node:stream';

async function* ghRepos() {
  const response = await fetch('https://api.github.com/users');
  const users = await response.json();

  for (let index = 0; index < 10; index++) {
    const reposResponse = await fetch(users[index].repos_url);
    yield await reposResponse.json();
  }
}

async function* ghFirsts() {
  for await (const repos of ghRepos()) {
    if (repos[0]) {
      yield repos[0].full_name + `\n`;
    }
  }
}

Readable.from(ghFirsts()).pipe(process.stdout);
```

> To better understand this example, look at *https://api.github.com/users* in your browser, find a user's repos_url, and look at that URL too.

The ghRepos function first fetches a list of users, then iteratively fetches (and then yields) the list of repos for the first 10 users. The ghFirsts function is then used to yield the name of the first repo from the list yielded by the ghRepos function.

Since the ghFirsts function is an async generator, we can use it to create a readable stream where each chunk is the name of the first repo for each user.

The pipeline method even supports having an async generator function in the mix. Instead of the last line, where we used .pipe, you can directly include ghFirsts within a pipeline call:

```
// Instead of:
// Readable.from(ghFirsts()).pipe(process.stdout);
```

```
await pipeline(ghFirsts(), process.stdout);
```

Figure 6-6 shows the output.

```
JS async-generator.js ✕

JS async-generator.js > ...
  1    import { pipeline } from 'node:stream/promises';
  2
  3  > async function* ghRepos() {…
 11    }
 12
 13  > async function* ghFirsts() {…
 19    }
 20
 21    await pipeline(ghFirsts(), process.stdout);
 22

 TERMINAL                                          🐚 bash  + ∨  ⬚  🗑  ⋯

● $ node async-generator.js
  mojombo/30daysoflaptops.github.io
  defunkt/ace
  pjhyett/auto_migrations
  wycats/abbot-from-scratch
  ezmobius/acl_system2
  ivey/advent-of-code-2022
  evanphx/ace
  vanpelt/amqp
  wayneeseguin/Adv360-Pro-ZMK
  brynary/active_admin
○ $
```

Figure 6-6. An async generator readable stream

The `pipeline` method also passes a stream object to an async generator that follows it in the pipeline, effectively enabling an async generator to act like a transform stream. Here's the uppercase transform example done with an async generator function:

```
import { pipeline } from 'node:stream/promises';

async function* upperCaseTr(source) {
  for await (const chunk of source) {
    yield String(chunk).toUpperCase();
  }
}

await pipeline(
  process.stdin,
  upperCaseTr,
  process.stdout
);
```

Readable streams can be consumed using an async iterator too:

```
for await (const chunk of readableStream) {
  console.log(chunk);
}
```

Streams Object Mode

By default, streams expect buffer or string values. There is an `objectMode` flag that we can set to have the stream accept any JavaScript object.

Here's a simple example to demonstrate that. The following combination of transform streams makes a feature to map a string of comma-separated values into a JavaScript object. So a,b,c,d becomes {a: b, c: d}:

```
import { Transform } from 'node:stream';

const commaSplitter = new Transform({
  readableObjectMode: true,

  transform(chunk, encoding, callback) {
    this.push(chunk.toString().trim().split(','));
    callback();
  },
});

const arrayToObject = new Transform({
  readableObjectMode: true,
  writableObjectMode: true,
  transform(chunk, encoding, callback) {
    const obj = {};
    for (let i = 0; i < chunk.length; i += 2) {
      obj[chunk[i]] = chunk[i + 1];
    }
    this.push(obj);
    callback();
  },
});

const objectToString = new Transform({
  writableObjectMode: true,
  transform(chunk, encoding, callback) {
    this.push(JSON.stringify(chunk) + '\n');
    callback();
  },
});
```

You can use these transform streams with the `pipeline` method:

```
await pipeline(
  process.stdin,
  commaSplitter,
```

```
        arrayToObject,
        objectToString,
        process.stdout,
    );
```

We pass an input string (for example, a,b,c,d) through commaSplitter, which pushes an array as its readable data (["a", "b", "c", "d"]). Adding the readableObjectMode flag on that stream is necessary because we're pushing an object there, not a string.

We then take the array and pipe it into the arrayToObject stream. We need a writableObjectMode flag to make that stream accept an object. It'll also push an object (the input array mapped into an object), and that's why we also need the read ableObjectMode flag there as well. The last objectToString stream accepts an object but pushes out a string, and that's why we need only a writableObjectMode flag there. The readable part is a normal string (the stringified object; see Figure 6-7).

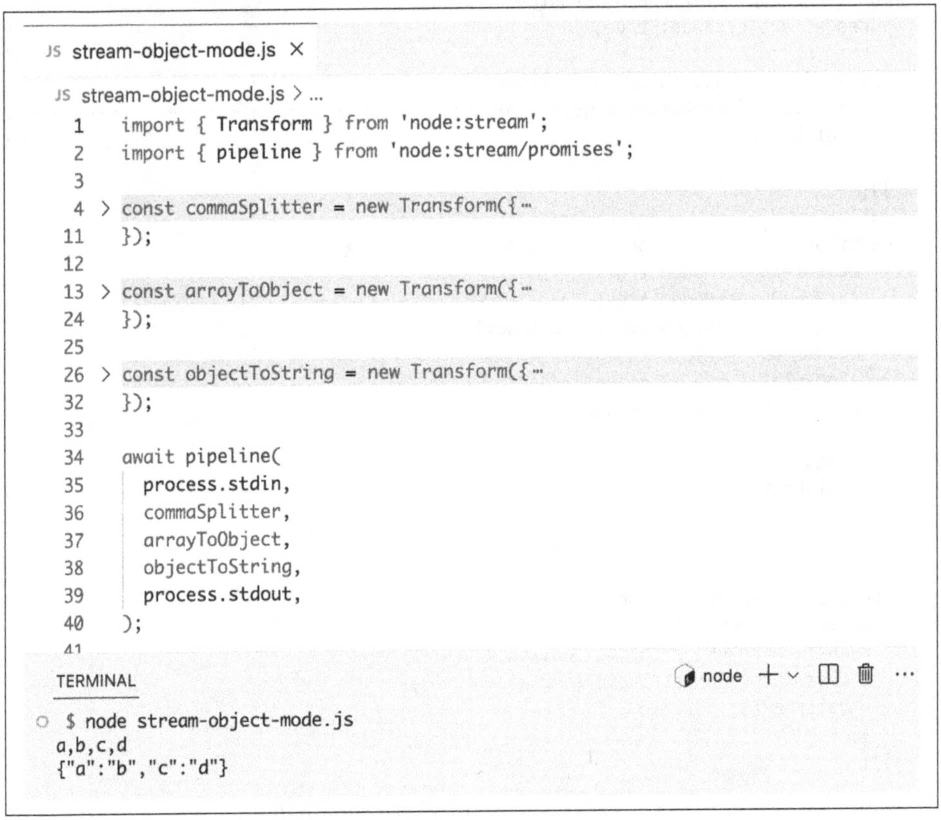

```
JS stream-object-mode.js ✕

JS stream-object-mode.js > ...
    1    import { Transform } from 'node:stream';
    2    import { pipeline } from 'node:stream/promises';
    3
    4  > const commaSplitter = new Transform({…
   11    });
   12
   13  > const arrayToObject = new Transform({…
   24    });
   25
   26  > const objectToString = new Transform({…
   32    });
   33
   34    await pipeline(
   35      process.stdin,
   36      commaSplitter,
   37      arrayToObject,
   38      objectToString,
   39      process.stdout,
   40    );
   41

TERMINAL                                          🍎 node + ∨ ⬛ 🗑 ⋯

○ $ node stream-object-mode.js
  a,b,c,d
  {"a":"b","c":"d"}
```

Figure 6-7. Stream object mode

Built-In Transform Streams

Node has a few very useful built-in transform streams such as the `zlib` and `crypto` streams.

Here's an example that uses the `zlib.createGzip()` stream combined with the `node:fs` readable/writable streams to create a file-compression script:

```
import fs from 'node:fs';
import zlib from 'node:zlib';
import { pipeline } from 'node:stream/promises';

const file = process.argv[2];

await pipeline(
  fs.createReadStream(file),
  zlib.createGzip(),
  fs.createWriteStream(file + '.gz')
);
```

You can use this script to gzip any file you pass its path as the argument. We're piping a readable stream for that file into the zlib built-in transform stream and then into a writable stream for the new gzipped file.

The cool thing about a pipeline of streams is that it enables us to compose our program piece by piece, in a readable way. Say, for example, you want the user to see a progress indicator while the file is being compressed. You can create a stream to report progress and make it part of the pipeline. An async generator function makes tasks like these easy:

```
await pipeline(
  fs.createReadStream(file),
  zlib.createGzip(),
  async function* (source) {
    for await (const chunk of source) {
      process.stdout.write('.');
      yield chunk;
    }
  },
  fs.createWriteStream(file + '.gz')
);
```

The applications of combining streams are endless. For example, if we need to encrypt the file before (or after) we gzip it, all we need to do is add another transform stream to the pipeline. The `node:crypto` module has the `createCipheriv` method that can be used for that purpose:

```
import crypto from 'node:crypto';

// ..
```

```
const algorithm = 'aes-256-ctr';
const key = crypto.randomBytes(32);
const iv = crypto.randomBytes(16);

await pipeline(
  fs.createReadStream(file),
  zlib.createGzip(),
  crypto.createCipheriv(algorithm, key, iv),
  fs.createWriteStream(file + '.gz'),
);
```

This code compresses and then encrypts the file it receives as an argument, which means the file can't even be unzipped with a normal unzip utility. Only someone who has the key/iv values can use this file. To actually be able to unzip anything zipped with this code, we need to use the opposite streams for `crypto` and `zlib` in a reverse order:

```
await pipeline(
  fs.createReadStream(file),
  crypto.createDecipheriv(algorithm, key, iv),
  zlib.createGunzip(),
  fs.createWriteStream(file.slice(0, -3)),
);
```

This code creates a read stream from the encrypted and zipped file, pipes it into the `crypto.createDecipheriv()` stream (using the same key/iv values), pipes the output of that into the `zlib.createGunzip()` stream, and then writes the content out back to a file without the `.gz` extension part.

Summary

Streams provide a memory-efficient way to process large data in Node. They allow us to work with data in chunks over time instead of needing to handle all the data at once.

The two main types of streams are readable streams (which produce data) and writable streams (which consume data). Duplex streams combine both functionalities, and transform streams are duplex streams that modify data in transit as well.

Streams have many methods and events that can be used together to ensure proper use. Readable streams can be in a flowing mode or a paused mode. A writable stream can be faster or slower than a readable stream. All these different states require the use of different methods and events. The easiest way to work with streams is through the use of the `pipeline` method as it automatically handles many of the complexities around using streams.

In the next chapter, we'll explore how to make a Node process execute a command from the OS and how to make a process fork another one.

CHAPTER 7

Child Processes

Running a Node instance in a single process works fine for a small application. As the application gets more complex and serves more users, a single process is not going to be enough to handle the increasing workload. No matter how powerful your server may be, a single thread can support only a limited load.

The fact that your Node code runs in a single thread does not mean that you can't take advantage of multiple processes and, of course, multiple machines as well.

Using multiple processes is the best way to scale a Node application. Node is designed for building distributed applications with many nodes. This is why it's named Node! Scalability is baked into the platform, and it's not something you start thinking about later in the lifetime of an application.

Introducing Child Processes

We can easily create a child process in a main Node process using the node:child_process built-in module. These child processes can communicate with each other with an event-driven messaging system because process objects implement the EventEmitter structure under the hood.

The node:child_process module enables us to efficiently run OS commands (like find, grep, etc.) from within a Node process, safely control the arguments passed to these commands, and do whatever we want with the commands' output. This gives a Node process a lot more power over what it can do.

In addition, the node:child_process module supports both buffers and streams to process the input and output of child processes. With streams, we can combine the power of multiple commands and run them together even if the task involves working with large amounts of data. We can simply pipe the output of one command as the input to another.

The node:child_process module offers four different ways to create a child process: spawn(), fork(), exec(), and execFile(). We're going to see the differences between these four functions and when to use each.

The spawn Function

The spawn function launches a command in a new process, and we can use it to pass that command any arguments. For example, here's code to spawn a new process that will execute the pwd Linux command:

```
import { spawn } from 'node:child_process';

const child = spawn('pwd');
```

We import the spawn function and execute it with the OS command as the first argument.

The result of executing the spawn function (the child object in this example) is a ChildProcess instance, which implements the EventEmitter API. This means we can register handlers for events on this child object directly. For example, we can do something when the child process exits by registering a handler for the exit event:

```
child.on('exit', function(code, signal) {
  console.log(
    `Child process exited. Code: ${code} - Signal: ${signal}`
  );
});
```

The handler function for an exit event gets the exit code and the signal that was used to terminate the child process. The signal variable is null when the child process exits normally.

Another important event to handle is the error event, which is emitted if the process could not be spawned (or terminated), or if it had any issues while executing. It's a good practice to always have a handler for this event:

```
child.on('error', (err) => {
  console.error(`Child process encountered an error: ${err.message}`);
});
```

The other events that we can register handlers for with the `ChildProcess` instances are `message`, `spawn`, `disconnect`, and `close`:

The `message` *event*

    Emitted when the child process uses the `process.send()` function to send messages. This is how parent/child processes can communicate with each other. We'll see an example of that shortly.

The `spawn` *event*

    Emitted once the child process has spawned successfully.

The `disconnect` *event*

    Emitted when the parent process manually calls the `child.disconnect` method.

The `close` *event*

    Emitted when the I/O streams of a child process are terminated. Every child process gets the three standard `stdio` streams, which we can access using `child.stdin`, `child.stdout`, and `child.stderr`. When these streams get closed, the child process that was using them emits the `close` event.

> This `close` event is different from the `exit` event because multiple child processes might share the same `stdio` streams, so one child process exiting does not mean that the streams got closed.

Since all streams are event emitters, we can listen to different events on the `stdio` streams that are attached to every child process. Unlike in a normal process, though, in a child process the `stdout`/`stderr` streams are readable streams while the `stdin` stream is a writable one. This is basically the inverse of those types found in a main process.

The events we can use for the `stdio` streams are the standard ones. Most importantly, on the readable streams we can listen to the `data` event, which will have the output of the command or any error encountered while executing the command:

```
child.stdout.on('data', data => {
  console.log(`child stdout:\n${data}`);
});

child.stderr.on('data', data => {
  console.error(`child stderr:\n${data}`);
});
```

The two handlers in this example will log both cases to the main process `stdout` and `stderr`. If we execute the code as it stands, the output of the `pwd` command gets printed, and the child process exits with code 0, which means no error occurred.

We can pass arguments to the command that's executed by the `spawn` function using the second argument of the `spawn` function, which is an array of all the arguments to be passed to the command. For example, to execute the `find` command on the current directory with a `-type f` argument (to list files only), we can do:

```
const child = spawn('find', ['.', '-type', 'f']);
```

If an error occurs during the execution of the command (for example, if we give the preceding `find` command an invalid destination), the `data` event handler on the `child.stderr` stream will be triggered, and the `exit` event handler on that instance will report an exit code of 1, which signifies that an error has occurred. The error values actually depend on the host OS and the type of error.

A child process `stdin` is a writable stream. We can use it to send a command some input. Just like any writable stream, the easiest way to consume it is using the `pipe` function. We simply pipe a readable stream into a writable stream. Since the main process `stdin` is a readable stream, we can pipe that into a child process `stdin` stream. Here's an example:

```
import { spawn } from 'node:child_process';

const child = spawn(`wc`);

process.stdin.pipe(child.stdin);

child.stdout.on(`data`, data => {
  console.log(`child stdout:\n${data}`);
});
```

In this example, the child process invokes the `wc` command, which counts lines, words, and characters in Linux. We then pipe the main process `stdin` (which is a readable stream) into the child process `stdin` (which is a writable stream). The result of this combination is that we get a standard input mode where we can type something, and when we hit Ctrl + D, what we type will be used as the input of the `wc` command.

We can also pipe the standard I/O of multiple processes on each other, just like we can do with Linux commands. For example, we can pipe the `stdout` of the `find` command to the `stdin` of the `wc` command to count all the files in the current directory:

```
import { spawn } from 'node:child_process';

const find = spawn('find', ['.', '-type', 'f']);
const wc = spawn('wc', ['-l']);
```

```
find.stdout.pipe(wc.stdin);

wc.stdout.on('data', data => {
  console.log(`Number of files ${data}`);
});
```

I added the -l argument to the wc command to make it count only the lines. When executed, the preceding code will output a count of all files in all directories under the current one.

Shell Syntax and the exec Function

By default, the spawn function does not create a *shell* to execute the command we pass to it. This makes it slightly more efficient than the exec function, which does create a shell. The exec function has one other major difference. It buffers the command's generated output and passes the whole output value to a callback function (instead of using streams, which is what spawn does).

Since the exec function uses a shell to execute the command, we can use any *shell syntax* directly with it. Shell syntax is the set of rules that guide how commands are structured and interpreted in a shell environment. For example, commands can have options and arguments.

There are also some special characters that we can use with commands. For example, the pipe symbol can be used to connect the output of one command to the input of another, the * character can be used in some commands to match files based on patterns, and the > character can be used to redirect the output of a command. You can even use loops and conditions and many other advanced features in shell commands.

Here's the previous find/wc example implemented with an exec function:

```
import { exec } from 'node:child_process';

exec('find . -type f | wc -l', (err, stdout, stderr) => {
  if (err) {
    console.error(`exec error: ${err}`);
    return;
  }

  console.log(`Number of files ${stdout}`);
});
```

Note how I used the shell native pipe feature through the pipe symbol (|) in this example.

Be careful when using the shell syntax in Node as it comes with a security risk, especially if you're executing any input that's provided externally. A user can simply initiate a command injection attack using shell syntax characters like ; (for example, command + '; rm -rf ~'). Strategies to mitigate that include sanitizing and validating input, making a list of acceptable commands, and restricting access control.

The exec function buffers the output and passes it to the callback function (the second argument to exec) as the stdout argument there. This stdout argument is the command's output that we want to print out.

The exec function is a good choice if you need to use the shell syntax and the size of the data expected from the command is small, since it buffers the whole data in memory before returning it.

The spawn function, with its stream objects, is a much better choice when the size of the output expected from the command is large. Using streams is compatible with the standard I/O objects that are often used with child processes. We can even make a spawned child process inherit the standard I/O objects of its parents if we want to. More importantly, we can make the spawn function use the shell syntax as well. Here's the same find | wc command implemented with the spawn function:

```
const child = spawn('find . -type f | wc -l', {
  stdio: 'inherit',
  shell: true
});
```

With the stdio: 'inherit' option, when we execute the code, the child process inherits the main process stdin, stdout, and stderr. This causes the child process data events handlers to be triggered on the main process.stdout stream, making the script output the result right away.

The shell: true option enables us to use the shell syntax in the executed command, just like we did with exec. But with this code, we still get the advantage of the streaming of data that the spawn function gives us. This is really the best of both worlds.

There are a few other good options we can use in the last argument to the node:child_process functions besides shell and stdio. For example, we can use the cwd option to change the working directory of the script. Here's the same count-all-files example done with a spawn function using a shell and with a working directory set to my *Downloads* folder:

```
const child = spawn('find . -type f | wc -l', {
  stdio: 'inherit',
  shell: true,
```

```
    cwd: '/Users/samer/Downloads'
});
```

Another option we can use is env to specify the environment variables that will be visible to the new child process. The default for this option is process.env, which gives any command access to the current process environment. If we want to override that behavior, we can simply pass an empty object as the env option or new values there to be considered as the only environment variables:

```
const child = spawn('echo $ANSWER', {
    stdio: 'inherit',
    shell: true,
    env: { ANSWER: 42 }
});
```

Since we specified an env object, the echo command will not have access to the parent process's environment variables. For example, it cannot access $HOME, but it can access $ANSWER because it was defined in the env property.

One last important child process option to explain here is the detached option, which makes the child process run independently of its parent process.

Assume we have a file, *timer.js*, that keeps the event loop busy:

```
setTimeout(() => {
    // Keep the event loop busy
}, 20_000);
```

We can execute *timer.js* in the background using the detached option:

```
import { spawn } from 'node:child_process';

const child = spawn('node', ['timer.js'], {
    detached: true,
    stdio: 'ignore'
});

child.unref();
```

The exact behavior of detached child processes depends on the OS. On Windows, they will have their own console window while on Linux they will be made the leaders of new process groups and sessions.

If the unref function is called on the detached process, the parent process can exit independently of the child. This can be useful if the child is executing a long-running process, but to keep it running in the background, the child's stdio configurations also have to be independent of the parent.

This example will run the *timer.js* script in the background by detaching and also ignoring its parent stdio file descriptors so that the parent can terminate while the child keeps running in the background.

The execFile Function

The `execFile` function behaves exactly like the `exec` function, but it does not use a shell, which makes it a bit more efficient and secure (because it has no shell syntax). We can use it to execute an external program or a script. Here's an example:

```
import { execFile } from 'node:child_process';

execFile('ruby', ['-e', 'puts rand(1..100)'], (error, stdout, stderr) => {
  if (error) {
    console.error(`Error: ${error}`);
    return;
  }

  console.log(`Output: ${stdout}`);
});
```

This example uses `execFile` to execute the `ruby` program, setting the arguments to execute a one-liner that'll print a random number between 1 and 100.

> On Windows, some files cannot be executed on their own, like *.bat* or *.cmd* files. Those files cannot be executed with `execFile`. Either `exec` or `spawn` with `shell` set to true is required to execute them.

Synchronous Child Process Functions

The functions `spawn`, `exec`, and `execFile` from the `node:child_process` module also have synchronous blocking versions that will wait until the child process exits:

```
import {
  spawnSync,
  execSync,
  execFileSync
} from 'node:child_process';
```

Those synchronous versions are potentially useful when trying to simplify scripting tasks or any startup processing tasks, but they should be avoided otherwise.

The fork Function

The `fork` function is a variation of the `spawn` function for spawning node processes. The biggest difference between `spawn` and `fork` is that a channel for interprocess communication (IPC) is established to the child process when using `fork`, so we can use the `send` function on the forked `process` along with the `process` object itself to

---

exchange messages between the parent and forked processes. We do this through the EventEmitter module interface. Here's an example.

In the parent file, *parent.js*, we have the following:

```
import { fork } from 'node:child_process';

const forked = fork('child.js');

forked.on('message', msg => {
  console.log('Message from child', msg);
});

forked.send({ hello: 'world' });
```

In the child file, *child.js*, we have the following:

```
process.on('message', msg => {
  console.log('Message from parent:', msg);
});

let counter = 0;

setInterval(() => {
  process.send({ counter: counter++ });
}, 1_000);
```

In the parent file, we fork *child.js* (which will execute the file with the node command), and then we listen for the message event. The message event will be emitted whenever the child uses process.send, which we're doing every second.

To pass down messages from the parent to the child, we can execute the send function on the forked object itself. Then, in the child script, we can listen to the message event on the process object.

When executing the preceding *parent.js* file, it'll first send down the { hello: 'world' } object to be printed by the forked child process. Then the forked child process will send an incremented counter value every second to be printed by the parent process.

Let's look at a more practical example of using the fork function. Let's say we have a web server that handles two endpoints. One of these endpoints (/compute) is computationally expensive and will take a few seconds to complete. We can use a long for loop to simulate that:

```
import { createServer } from 'node:http';

const longComputation = () => {
  let sum = 0;
  for (let i = 0; i < 1e9; i++) {
    sum += i;
```

```
  }
  return sum;
};

const server = createServer();

server.on('request', (req, res) => {
  if (req.url === '/compute') {
    const sum = longComputation();
    return res.end(`Sum is ${sum}`);
  } else {
    res.end('Ok');
  }
});

server.listen(3000, () => {
  console.log('Server is running...');
});
```

This program has a big problem; when the /compute endpoint is requested, the server will not be able to handle any other requests because the main thread is busy with the long for loop operation.

There are a few ways we can solve this problem depending on the nature of the long operation, but one solution that works for all cases is to just move the computational operation into another process using the fork function.

We first move the whole longComputation function into its own file and make it invoke that function when instructed via a message event from the main process. In a new *compute.js* file:

```
const longComputation = () => {
  let sum = 0;
  for (let i = 0; i < 1e9; i++) {
    sum += i;
  }
  return sum;
};

process.on('message', msg => {
  const sum = longComputation();
  process.send(sum);
});
```

Now, instead of doing the long operation in the main thread, we can fork the *compute.js* file and use the messages interface to communicate messages between the server and the forked process:

```
import { createServer } from 'node:http';
import { fork } from 'node:child_process';

const server = createServer();
```

```
server.on('request', (req, res) => {
  if (req.url === '/compute') {
    const compute = fork('compute.js');
    compute.send('start');
    compute.on('message', sum => {
      res.end(`Sum is ${sum}`);
    });
  } else {
    res.end('Ok');
  }
});

server.listen(3000, () => {
  console.log('Server is running...');
});
```

When a request to /compute happens now with this code, we simply send a message to the forked process to start executing the long operation. The main thread will not be blocked.

Once the forked process is done with that long operation, it can send its result back to the parent process using process.send.

In the parent process, we listen to the message event on the forked child process itself. When we get that event, we'll have a sum value ready for us to send to the requesting user over HTTP.

This method is limited by the number of processes we can fork, but when we execute it and request the long computation endpoint over HTTP, the main server is not blocked at all and can take further requests.

Node's cluster module, which is the topic of Chapter 9, is based on this idea of child-process forking and load balancing the requests among the many forks that we can create on any system.

Summary

We can create child processes in Node to scale applications beyond a single thread and access external functionalities from the OS and other programs.

Node child processes enable us to execute external shell commands, programs, and scripts. We can also establish communication between child processes and their parent process when we need to.

The child_process module in Node has various flexible asynchronous functions like spawn, exec, and execFile and matching synchronous versions that will wait until the child process exits. The spawn function can be used to stream the output of execution, making it suitable for handling large outputs. Shell syntax can be used in spawn and exec.

The fork function can be used to delegate running a Node script to a different process and free the main process to do more.

Child process objects are event emitters. They emit events like exit, error, and message. The message event is how a child process can communicate with its parent process.

In the next chapter, we'll learn about the Node's built-in modules for writing tests and assertions.

CHAPTER 8

Testing Node

Testing in the real world is generally about finding what's wrong with something. When it comes to code, testing is about finding what's right. It's the guarantee that something is working.

Untested code is simply dangerous. Efficient and scalable code is well-tested. Testing is not only about making sure the code is doing what it is supposed to do; it's also about making sure the code continues to do what it is supposed to do after any changes in the code, its environment, and its use patterns. Testing is about writing high-quality code and catching potential problems as early as possible.

Regular testing of code keeps it healthy and makes maintaining it easier. It also increases the confidence of its maintainers to make changes. Making changes to untested code is a recipe for disaster. A new feature in module X might break other features in module Y. You can't keep the dependencies in your head. You can't test all the code manually every time there is a change. There is no way around it. You have to write code to test your code, and yes, your testing code might have problems too. Tests might introduce false negatives and false positives. That's why it's extremely important to get the tests right. That is what this chapter is all about.

Assertions and Runners

A test in Node is a set of *assertions*. To execute the tests, you need a test *runner*. Node has built-in modules for both. Here's a simple test for the map method on arrays:

```
import test from 'node:test';
import assert from 'node:assert/strict';
```

```
test('doubles items for a 2*e transformer', () => {
  // Arrange
  const inputArray = [1, 2, 3, 10];
  const expectedResult = [2, 4, 6, 20];

  // Act
  const actualResult = inputArray.map((e) => 2 * e);

  // Assert
  assert.deepEqual(actualResult, expectedResult);
});
```

It helps to think of tests with the *Arrange-Act-Assert pattern*:

Arrange

> This section is the setup. It's where we prepare the environment and prerequisites of the test.

Act

> This section is where we execute the method we want to test.

Assert

> This section is where we confirm our expectations.

To execute this test, you just put it in a Node script and run it with the node command.

Figure 8-1 shows the output.

The node:test module has a few objects that can be used to organize your tests and label them. I used the test method in this example to give the test the name doubles items for a 2*e transformer. When the node command sees this method, the built-in runner carries out the assertions in it to produce this output, which lists all test names and whether they succeeded or failed, along with the time duration for each and a diagnostic summary about all tests. As you can see from this example, the one test we did was OK.

```
4    test('doubles items for a 2*e transformer', () => {
5      // Arrange
6      const inputArray = [1, 2, 3, 10];
7      const expectedResult = [2, 4, 6, 20];
8
9      // Act
10      const actualResult = inputArray.map((e) => 2 * e);
11
12      // Assert
13      assert.deepEqual(actualResult, expectedResult);
14    });
15
```

```
TERMINAL                                        bash + ∨  ⬚  🗑  ⋯

● $ node map.test.js
  ✓ doubles items for a 2*e transformer (1.201458ms)
  ℹ tests 1
  ℹ suites 0
  ℹ pass 1
  ℹ fail 0
  ℹ cancelled 0
  ℹ skipped 0
  ℹ todo 0
  ℹ duration_ms 14.01825
○ $ |
```

Figure 8-1. A passing test

The node:assert module has a set of assertion functions that we can use to implement the *Assert* section of any test. In this example, I used the deepEqual method to make sure the actual mapped array is deeply equal to the expected array. The simple assert.equal method will not work here as the arrays are different objects. The deep Equal method checks the equality of the array items, which is what we want here.

> Note how I used node:assert/strict in this example and that's what you should always do. The legacy mode for this module (without the /strict) is based on the == operator (which performs type conversions of the operands before comparison) and should be avoided.

Instead of the simple `test` method, we can use the `describe` and `it` methods:

```
import { describe, it } from 'node:test';
import assert from 'node:assert/strict';

describe('The map method on arrays', () => {
  it('doubles items for a 2*e transformer', () => {
    const inputArray = [1, 2, 3, 10];
    const expectedResult = [2, 4, 6, 20];

    const actualResult = inputArray.map((e) => 2 * e);

    assert.deepEqual(actualResult, expectedResult);
  });
});
```

The `it` method is basically an alias to `test`, but I prefer it as a reminder to stay consistent on labeling tests by answering the question, "What does this test validate?" and finishing the sentence: "It ___."

The `describe` method gives us a way to group tests together and describe their purpose. We'll usually have multiple tests written for a method like `map`. We use `describe` to group them. This affects the output of running the tests as well.

Deep Versus Shallow Equality

When we check a `===` b, that's a shallow equality check that's making sure both a and b refer to the same memory location.

Deep equality, on the other hand, is when two objects have the same structure and values, regardless of their memory location.

For example, consider the following two arrays:

```
const arrA = [1, 4, 16];
const arrB = [2-1, 2*2, 4**2];

assert.equal(arrA, arrB);       // AssertionError
assert.deepEqual(arrB, arrB);   // Ok
```

Since the two arrays are different objects, the `assert.equal` check will fail. However, since both arrays have three elements and each element in `arrA` equals the element in `arrB` that's in the same position, these two arrays are deeply equal, and the `assert.deepEqual` check will pass.

Let's see what a failed test run looks like. Replace `deepEqual` with `equal` in the map method test, and run it again. Figure 8-2 shows the output I get.

```
 7      const expectedResult = [2, 4, 6, 20];
 8
 9      const actualResult = inputArray.map((e) => 2 * e);
10
11      assert.equal(actualResult, expectedResult);
12    });
13  });
```

TERMINAL   bash + ∨ ⊡ 🗑 …

```
$ node map.test.js

▶ The map method on arrays (5.967083ms)
ℹ tests 1
ℹ suites 1
ℹ pass 0
ℹ fail 1
ℹ cancelled 0
ℹ skipped 0
ℹ todo 0
ℹ duration_ms 28.350917

✖ failing tests:

test at map.test.js:5:3
✖ doubles items for a 2*e transformer (4.334292ms)
  AssertionError [ERR_ASSERTION]: Values have same structure but are not reference-
equal:

  [
    2,
    4,
    6,
    20
  ]
```

Figure 8-2. A failing test

Since the expected array does not equal the actual array, using `assert.equal` fails here.

> I used the `map` method as an example here, but we shouldn't really be testing standard library methods like that. They are already well-tested. We should be testing our own code, our functions, our modules, our dependencies, and our systems. To start on that, let's talk about the different types of tests.

Types of Tests

There are four major types of tests: *unit, functional, integration,* and *end-to-end.* Each has different scopes and purposes. Practically, some of these types overlap sometimes, and in certain cases, some tests can be categorized under multiple types. However, I think learning the difference between these types makes you write better tests.

To understand the difference between these types, let's start with a code example and see how different tests can be written for it. Let's assume we have two modules to manage products and orders in a database. For simplicity, we'll use simple arrays to store the records.

In a *products.js* file, we have the following:

```
let products = [
  { id: 1, name: 'Phone', price: 600 },
  { id: 2, name: 'Laptop', price: 2000 },
  { id: 3, name: 'Headphone', price: 100 },
];

const addProduct = (product) => {
  products.push(product);
  return product;
};

const getProductById = (id) => {
  return products.find((product) => product.id === id);
};

export { addProduct, getProductById };
```

In an *orders.js* file, we have the following:

```
const orders = [];

const createOrder = (productId, quantity) => {
  const order = { productId, quantity, status: 'pending' };
  orders.push(order);
  return order;
};

const updateOrderStatus = (orderId, status) => {
  const order = orders.find((order) => order.id === orderId);
  if (order) {
    order.status = status;
  }
  return order;
};

export { createOrder, updateOrderStatus };
```

I initialized the products array with some testing data. Testing data is usually seeded for the tests in a different phase—for example, a setup phase run before all the tests. For simplicity, I included it in the module directly.

Unit Tests

A *unit test* is written for a small part of the code (a *unit*). A unit could be a single function, a group of related functions, a module, a component, or anything else that can be tested on its own, in isolation. Usually, any external dependencies are faked with double objects to maintain testing a unit in isolation. More on that shortly.

An example of a unit test for these modules is one that tests a single function in either module. For example, we should test the `getProductById` against one of the known testing data products and against a product that does not exist:

```
import { describe, it } from 'node:test';
import assert from 'node:assert/strict';

import { getProductById } from './products.js';

describe('getProductById', () => {
  it('finds a product that exists', () => {
    const product = getProductById(2);
    assert.deepEqual(product, {
      id: 2,
      name: 'Laptop',
      price: 2000,
    });
  });

  it('returns undefined for a product that does not exist', () => {
    const product = getProductById(-1);
    assert.equal(product, undefined);
  });
});
```

For the remaining examples, I'll omit the `import` statements for `node:test` and `node:assert` for brevity.

Functional Tests

A *functional test* is written for a *feature* of the code (a functionality). A feature could be a single unit or multiple units, but this type of testing is not run in isolation. Functional tests often interact with external dependencies like databases and networks.

An example of a functional test for the orders module is one that tests the functionality of placing an order and updating its status:

```
import { createOrder, updateOrderStatus } from './orders.js';

describe('Order Management', () => {
  it('places an order and updates its status', () => {
    const newOrder = createOrder(1, 1);
    const updatedOrder = updateOrderStatus(
      newOrder.id,
      'completed',
    );
    assert.equal(updatedOrder.status, 'completed');
  });
});
```

Integration Tests

An *integration test* is written to check that different modules or services work correctly together. While functional tests are testing the functionality of a single module or service, integration tests are used to check how multiple modules or services integrate with each other.

An example of an integration test for the products/orders modules is one that tests how order creation integrates with product retrieval:

```
import { getProductById } from './products.js';
import { createOrder } from './orders.js';

describe('Order Creation', () => {
  it('integrates with product retrieval', () => {
    const product = getProductById(1);
    const order = createOrder(product.id, 2);
    assert.equal(order.productId, product.id);
    assert.equal(order.quantity, 2);
  });
});
```

End-to-End Tests

An *end-to-end (e2e) test* is written to simulate a real-world usage of your code, from start to finish, across all components, modules, and services.

An example of an e2e test for the products/orders modules is one that tests the entire flow from product creation to order completion:

```
import { addProduct, getProductById } from './products.js';
import { createOrder, updateOrderStatus } from './orders.js';

describe('From product addition to order completion', () => {
  it('works', () => {
```

```
      addProduct({ id: 4, name: 'Tablet', price: 500 });
      const product = getProductById(4);
      const order = createOrder(product.id, 1);
      const finalOrder = updateOrderStatus(order.id, 'completed');
      assert.equal(product.name, 'Tablet');
      assert.equal(finalOrder.status, 'completed');
    });
  });
```

Knowing the different types of tests is good in general. We get to focus our testing strategy based on the needs of the application, and perhaps more importantly, when a test fails, its type is the first indicator of what went wrong.

However, there's often a thin line between these different types of tests. For example, to better unit-test the `getProductById` function, we should have the test add a product first, then try to find it. You might think of that as a kind of integration testing. The labels are not really that important. What's important is having a strategy that suits the application and organizing testing code according to that strategy. In an application with isolated modules that don't depend on each other at all, there's little place for integration tests. The focus in that application should be on unit tests. For other types of applications, the team might opt to not write unit tests at all and rely completely on integration tests and e2e tests.

Test Doubles

In many cases, certain behaviors need to be simulated for testing purposes. For example, when testing code that uses an external API, it's a bad idea to have your tests hit the actual API every time you run them. It's better to use a test double object to fake the response of the API and regularly maintain this test double based on actual responses from the API.

Test doubles are basically objects we use to simulate the behavior of real objects. Just like a stunt double in a movie performs actions in place of a real actor, a test double object simulates the action of a real object.

Test doubles are commonly used in unit tests, as these tests are usually focused on isolated units. For example, if you need to unit-test a function that first retrieves a record from a database, you can use a test double for the database retriever object and focus the test on the remaining logic of that function. An integration or e2e test can take care of testing the actual retrieval of records from the database.

Using test doubles simplifies testing and improves the speed of running all tests. With the right test doubles, you don't need to set up a database, an API, or network channels to run simple tests in isolation.

There are a few different types of test doubles. To see them in action, let's start with a code example:

```
function notifyCompletion(taskId, { database, emails, mailer }) {
  const status = database.getStatus(taskId);

  if (status === 'complete') {
    const email = emails.notifications;
    mailer.sendEmail(email, `Task ${taskId} is complete.`);
  }
}
```

This is a simple function that has three dependencies: a `database` object, an `emails` object, and a `mailer` object. All three dependencies are injected into the function as arguments (following the dependency injection pattern).

Different types of test doubles can be used to simulate these services with varying features for testing.

Dummy objects are passed around but are never actually used. They are just placeholders. They do nothing at all.

For example, if we're not focusing on the `emails` service, we can use a dummy empty object in a test:

```
// Assert something
notifyCompletion(123, {
  database,
  emails: {},
  mailer,
});
```

Stub objects are used to replace a unit of code with one that always behaves in a fixed way. To use a stub for a function, we write a new function that will always return the same result.

For example, to stub the `database` service in `notifyCompletion`, we can do something like this:

```
const stubDatabase = {
  getStatus: function (taskId) {
    return 'complete';
  },
};

// Assert something
notifyCompletion(123, {
  database: stubDatabase,
  emails,
  mailer,
});
```

Mock objects are used to test interactions and flows between components. We can use them for cases when we want to make sure that a certain function is called with certain arguments, a function tries to send an email, or a set of methods are called in a specific order.

For the `notifyCompletion` case, we can mock the `mailer` object to make sure its `sendEmail` method is called with the right arguments:

```
const mockMailer = {
  sendEmail(email, message) {
    assert.equal(email, 'task@example.com');
    assert.equal(message, 'Task 123 is complete.');
  },
};

// Assert something
notifyCompletion(123, {
  database,
  emails,
  mailer: mockMailer,
});
```

Spy objects are used to gather information about function calls, such as how many times a function is called.

If we're interested in keeping track of how many times the `sendEmail` method is used, we can use the following spy object for it:

```
const spyMailer = {
  sendCount: 0,
  sendEmail (email, message) {
    this.sendCount++;
  },
};
```

Fake objects are basically a simplified implementation of a complex interface or class. They work like the real code but with some complexities removed.

For example, if we need to test both the `complete` and `incomplete` values for `get Status`, we can use a fake `database` object like the following:

```
class FakeDatabase {
  constructor() {
    this.tasks = { 123: 'incomplete', 456: 'complete' };
  }

  getStatus(taskId) {
    return this.tasks[taskId];
  }
}
```

As you can see, choosing what test doubles to use completely depends on what you're testing and how much of it you want to test.

Test doubles don't have to be purely one of these types. We can combine them. We can make FakeDatabase spy on how many times getStatus is called and throw in some expectations to make it also a mock double.

The names and scopes of these types are good to know and understand, but don't be overwhelmed with them. Just remember that a test double object can be used to simulate the behavior of a real object, and it adds power to your test. You can use them to simplify your tests, keep track of calls, and verify expectations too.

Node has a built-in mock object within the node:test module that can be used to create test doubles and has some powerful features as well. Let's talk about a few examples.

You can use Node's mock object to mock a method on any object, including methods within built-in modules. For example, if you're testing a function that uses the read File method from Node's node:fs module, you can use the mock object to skip the actual reading of files for testing purposes:

```
import fs from 'node:fs/promises';
import { mock } from 'node:test'

mock.method(fs, 'readFile', async () => 'Hello World');
```

Once a method is mocked, it gets a .mock property that can be used to track how and when that method is called. The following assertion makes sure the readFile method is called exactly one time:

```
assert.equal(
  fs.readFile.mock.calls.length,
  1,
);
```

The mock object can be used to mock timer functions like setTimeout and set Interval:

```
// Create a mock function
const fn = mock.fn();

// Mock the setTimeout API
mock.timers.enable({ apis: ['setTimeout'] });
// Now, any calls to setTimeout will not
// create a timer.

// Test it
setTimeout(fn, 500);
assert.equal(fn.mock.callCount(), 0);

// You can manually tick the mocked timer
```

```
mock.timers.tick(500);
assert.equal(fn.mock.callCount(), 1);
```

You can also use the `mock` object to mock date objects when you need to write a time-dependent test:

```
test('mocks the Date object', (context) => {
  // Mock the Date object
  context.mock.timers.enable({ apis: ['Date'] });

  // The initial date will be based on 0 in the UNIX epoch
  assert.equal(Date.now(), 0);

  // Advance in time will also advance the date
  context.mock.timers.tick(100);
  assert.equal(Date.now(), 100);
});
```

Note how in this example I used the test `context` argument. This is passed to each test function and can be used to interact with the test runner. It has many of the `node:test` module features attached to it for convenience. The `context.mock` object is one of them. You can use it directly without importing `mock` from `node:test`. Some of the context methods have advantages over their noncontext matching methods. For example, by using `context.mock`, the test runner will automatically restore all mocked functionality once the test finishes. This is not true for using the `mock` object directly; in fact, with a direct `mock` object use, we'll need to manually reset things after we're done with them using a `mock.reset()` call.

There are many other methods attached to the context object. Here are a few examples:

`context.test`
: This method can be used to structure your tests in a hierarchical manner by nesting subtests under a top-level test.

`context.diagnostic`
: This method can be used to write a message to the test runner output. Any diagnostic information is included at the end of the test's results.

`.before`, `.after`, `.beforeEach`, *and* `.afterEach`
: These methods can be used to execute code before/after all tests in a scope and before/after each test.

Here is an example to demonstrate:

```
test('top level test', async (t) => {
  t.afterEach((t) => t.diagnostic(`Finished running ${t.name}`));
  t.after((t) => t.diagnostic(`Finished running ${t.name}`));

  await t.test('subtest 1', (t) => {
```

```
      assert.equal(1, 1);
    });

    await t.test('subtest 2', (t) => {
      assert.equal(2, 2);
    });
  });
```

Figure 8-3 shows the output.

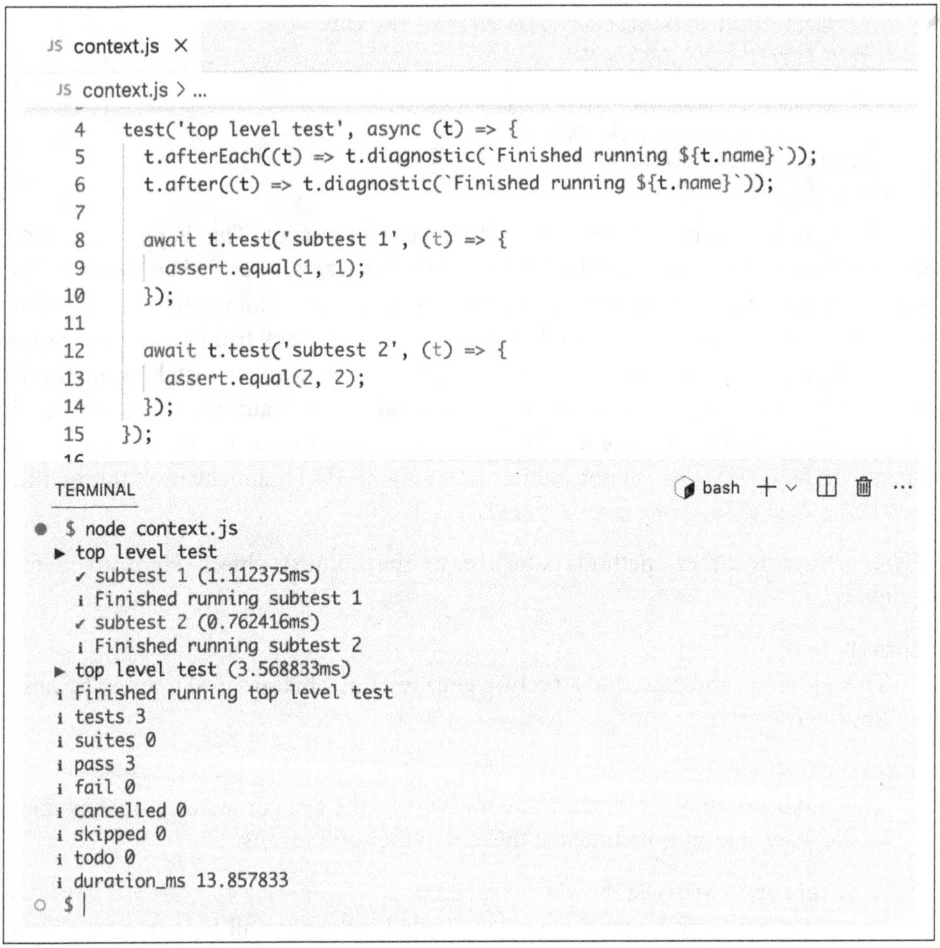

Figure 8-3. The example test with context

Note how ${t.name} inside afterEach refers to the subtests, whereas it refers to the top-level test inside after.

The after, before, afterEach, and beforeEach functions are also available as imports from node:test.

Organizing and Filtering Tests

Where should test code be placed in an application? The two main strategies are to either place test files right next to the code files they test or isolate them in folders that can be mapped to the code they test.

For the first strategy, you put a test file right next to the module it's testing. If your module is *orders.js*, you create *orders.test.js* (or *ordersTest.js*, or *orders.spec.js*) in the same folder as *orders.js*. Pick one of the naming formats, and stick with it throughout the entire application.

The main advantage of this approach is to quickly find the tests for any code you're dealing with and to quickly figure out when something isn't tested at all. I think this strategy is great for unit and functional tests.

For the other strategy, you mirror the structure of the application into a *tests* directory. For example, if you have *app/domain* and *app/models* folders, you would create *tests/domain* and *tests/models* directories.

The main advantage of this approach is to keep the code and tests separate while still being able to quickly find tests. However, this dependency on the application structure is not ideal. It ties the tests to any problems in the structure, and when a restructure is needed, a match of that restructure needs to happen in the *tests* folder as well, because otherwise it'll get harder to find tests.

A variant of this strategy is to maintain a *tests* folder within each *app* folder. You create *app/domain/tests* and *app/models/tests* folders and put all tests related to these modules in there. This still follows the app structure, maintains isolation between code and tests, and is easier to deal with when things get restructured.

No matter what approach you pick for your tests, consistency is the key. Pick a folder-naming convention and a file-naming convention as well, and stick with them for the entire application.

> You can use code quality tools like ESLint and its plugins to flag any divergence from the naming conventions of a project.

Consistency in test organizations helps when you need to run a subset of tests. As you're working on one part of an application, you would usually frequently run the tests related to that part. You would still need to run all the tests, but you can do that less frequently.

Node supports a few ways for you to filter what tests to run. A second optional argument to the main methods of `node:test` can be used to skip a test altogether, mark it as a TODO test (which is executed but not included in failed tests), or mark a test as the only test that should be executed:

```
// The following test will be skipped
test('something', { skip: true }, () => {
  // ...
});

// The following test will be marked as TODO
test('something', { todo: true }, () => {
  // ...
});

// The following test will be the only executed test
// when the `node` command is run
// with a `--test-only` option
test('something', { only: true }, () => {
  // ...
});
```

You can also filter tests by their names. Node has two options for that: `--test-name-pattern="something"` and `--test-skip-pattern="something"`. This gives you the option of running or skipping tests when their names match certain patterns.

Combining all these features gives you a lot of flexibility in running and maintaining test files.

Test-Driven Development

Instead of writing tests for code, why not write code for tests? This is the test-driven development (TDD) approach. You start with a test, which will fail at first since there is no code. You write the code to make that test pass. Then you write another failing test, make it pass with more code, and so on. This creates a cycle of red-green (or fail-pass for tests). While you're in a green state, you can refactor and improve your code, which is why this cycle is known as the red-green-refactor cycle.

The TDD inverted cycle might feel weird, but it's actually incredibly powerful. If you strictly follow a TDD approach, you will simply not have any code that's not tested. You can only write code to make a red test green, after all.

Not only that, I think TDD forces you to write better code. You don't waste time on unnecessary code. For any piece of code you're thinking of, you're forced to think of a test case for it. It gives you a clear understanding of the code and helps you make better decisions about how the code should evolve.

While TDD is simple to use, there are a few things to keep in mind. Let's look at an example. Let's say we want to create a function that validates email addresses. To use TDD, we might start with a test case like this:

```
describe('validateEmail', () => {
  it('works for a normal email', () => {
    assert(validateEmail('test@example.com'));
  });
});
```

Then we implement the simplest code that can make this test pass. This is important. Don't attempt to write more code than the absolute minimum of what can make this test pass. After you make the test pass, you can add another test case:

```
it('works for less common emails', () => {
  assert(validateEmail('test.one@example.com.ab'));
  assert(validateEmail('test+one@1.com.ab'));
  assert(validateEmail('123@a-z.com.ab'));
});
```

Write the code to make the test pass, then add another test:

```
it('works for non-english characters', () => {
  assert(validateEmail('test@mañana.org'));
});
```

Write the code to make the test pass, then add another test:

```
it('fails for invalid addresses', () => {
  assert(validateEmail('test@') === false);
  assert(validateEmail('@test.com') === false);
});
```

Note how the thinking is explicit here with tests, and every thought we have is documented with a test. For the last case, you would start thinking: should the function return false or should it throw an error? You make these decisions while designing the tests.

> TDD is great for unit tests and some functional tests, but it gets more challenging for bigger tests.

Continuous Integration

When working in a Node project that has tests, you should continuously run tests while making changes in your development environment. To make this easier, you can use the organizing and filtering techniques discussed earlier. You can also use the `--watch` option to automatically run tests every time there is a change.

This process is known as *continuous testing*. You should not push any code without making sure all tests pass locally first. But even tests that pass for you locally might not pass in another testing environment. This is why a tested project needs a continuous integration (CI) strategy as well.

With CI, you can use a platform like GitHub Actions or GitLab CI/CD to force a run for all tests after new changes are proposed to the project (for example, in a pull request). Changes cannot be merged to the project unless the tests pass first. CI tools run all the tests in the cloud, and they can be integrated with your source control tools. For example, you can make them automatically block a pull request whose changes made the test suite fail.

CI leaves no room for common mistakes. Maybe someone did not run all the tests while making a change, or maybe the tests passed for them in their local development environment but would not pass in an environment closer to production. CI is the fail-safe approach.

Another subtle advantage of having a CI pipeline is that history can be preserved to track the progress of a suite of tests. You can track things like the time it takes for all the tests to run and the coverage of these tests, and you can also track how these things evolve.

Coverage is an indicator of how much of the code is tested. It's a good indicator for quantity, but it has nothing to do with quality. However, ensuring a good coverage amount is a good first step. You can set up your CI pipeline to generate a coverage report (the Node test runner has some support for that) and then demand that the percentage of code covered with tests not go below a certain threshold. If you push untested code, the coverage will decrease, and your code will not be merged.

Summary

While a small and simple application might do OK without tests, as the code evolves, a good testing strategy becomes the only way to grow an application with confidence.

Node has built-in tools for testing. There's the `node:test` module that provides methods to structure and run tests and the `node:assert` module that provides methods to perform assertions in tests.

There are four main types of tests: unit tests are used to test a single component in isolation, functional tests are used to test one functionality, integration tests are used to test the integration between multiple parts of an application, and end-to-end tests are used to test a complete interaction with the application, from start to finish.

You can use test double objects (mocks, stubs, spies, etc.) to replace certain parts of the code for the purpose of isolating what to test and improving test quality and speed.

Test files can be collocated with the code they test or organized in a structure that mirrors the application structure. When running tests, you can have filters to run a subset of the tests and focus on only that subset.

In a well-tested Node application, you should integrate tests in your development workflow. You should run all tests locally before you push a change, or even better, write and run the tests before and after you make a change (which is known as test-driven development). You should also have the tests run in a staging environment through a continuous integration pipeline that gets automatically triggered for every new change that's proposed to the project, and use it to block that change if it makes any tests fail.

In the next chapter, we'll explore Node's cluster module and see how it can be used to manage running multiple Node processes.

CHAPTER 9

Scaling Node

Scaling an application is about making it able to handle more work without slowing down or crashing. You can scale any application by either giving its servers more memory or CPU power, or by adding more servers.

For Node, with its non-blocking and event-driven model, scaling is something that's baked into the core of the runtime environment. A Node application should be composed of multiple small, distributed nodes. You can run the same Node process on multiple CPU cores (or multiple servers) and then load balance the requests among them. Node has a built-in module to help with that.

In this chapter, we'll learn all about the Node `node:cluster` module, which can help improve the workload performance of a Node process by making it utilize the full CPU power of a server. It also improves the availability and uptime of servers.

Strategies of Scalability

While making an application able to handle more work is the most popular reason to scale them, there are more reasons. Applications are also scaled to increase their availability and their tolerance to failure.

If you're tasked with scaling an existing application, your options are either to scale it *vertically* by giving its servers more power (memory and CPU) or *horizontally* by adding more servers. However, if you're thinking of scaling as you build your applications (which you should be), you need to understand the following three strategies and make scalability choices based on that understanding and the applications' projected use over time:

Cloning

The easiest thing to do to scale a big application is to clone it multiple times and have each cloned instance handle part of the workload (with a load balancer, for example). This does not cost a lot in terms of development time, and it's highly effective. This strategy is the minimum you should do, and Node has the built-in `node:cluster` module to make it easier for you to implement the cloning strategy on a single server.

Decomposing

We can also scale an application by decomposing it based on functionalities and services. This means having multiple applications with different code bases and sometimes with their own dedicated databases and UIs. This strategy is commonly associated with the term *microservice*, where *micro* indicates that those services should be as small as possible (in reality, the size of the service is not what's important but rather the enforcement of loose coupling and high cohesion between services). The implementation of this strategy is often not easy and could result in long-term unexpected problems, but when done right the advantages are great.

Splitting

We can also split the application into multiple instances where each instance is responsible for only one part of the application's data. This strategy is often named *horizontal partitioning*, or *sharding* in databases. Data partitioning requires a lookup step before each operation to determine which instance of the application to use. For example, maybe we want to partition our users based on their country or language. We need to do a lookup of that information first.

Applications might pick one or more of these strategies based on their current and projected needs.

The Cluster Module

Node's `node:cluster` module can be used to run multiple instances of the same Node process on different CPU cores and balance the working load among them. It uses the `child_process.fork` method under the hood to give us an easy way to fork an application process as many times as we need. On one server machine, we can fork a process as many times as there are CPU cores on that machine. The `node:cluster` module can then take over the management of these forks, load balance all requests coming to the application across all forked processes, and reload any forks when needed.

The `node:cluster` module is Node's helper for us to implement the cloning scalability strategy vertically on one server. When you have a big server with a lot of resources or when it's easier and cheaper to add more resources to one server rather than adding new servers, the node:cluster module is a great option for a really quick implementation of the cloning strategy.

Even small servers usually have multiple cores, and even if you're not worried about the load on your Node server, you should use the node:cluster module anyway to increase your server availability and fault tolerance. It's a simple step with big advantages.

Taking advantage of what the node:cluster module has to offer is simple. You create a *primary* process that forks a number of *worker* processes and manages them. Each worker process represents an instance of the application. All incoming requests are handled by the primary process, which is the one that decides which worker process should handle an incoming request.

The primary process's job is simple. It actually just uses a round-robin algorithm to pick a worker process. This is enabled by default on most platforms, but it can be modified to let the load balancing be handled differently (for example, by the OS or with custom logic).

The round-robin algorithm distributes the load evenly across all available worker processes on a rotational basis. The first request is forwarded to the first worker process, the second to the next worker process, and so on. When the end of the worker process list is reached, the algorithm starts again from the beginning.

Round-robin is one of the simplest load-balancing algorithms, but there are several other algorithms that can be used for load balancing:

Least connections
> The primary process sends an incoming request to the worker with the fewest active connections. This helps distribute the workload more evenly.

Weighted round-robin
> Workers are assigned different weights, and those with higher weights handle more incoming requests.

Random
> The primary process picks a worker randomly to handle an incoming request.

IP hash
> The primary process uses the IP of the client making a request to determine which worker should handle that request.

URL hash
> The primary process uses the requested URL to determine which workers should handle that request.

Each routing algorithm has its own use cases, benefits, and disadvantages. Picking an algorithm depends on the requirements of the application and the environments in which it runs.

Primary and Worker Processes

To see the `node:cluster` module in action, let's use this basic HTTP server. Create a *slow-server.js* file for this code:

```
import { createServer } from 'node:http';

createServer((req, res) => {
  for (let i = 0; i < 1e8; i++) {
    // Simulate CPU work:
  };

  res.end();
}).listen(3000, () => {
  console.log(`Process ${process.pid}`);
});
```

Note how I added an empty `for` loop to simulate some CPU work before responding.

With the `node:cluster` module, we'll be working with multiple processes, so I'm logging the process ID (with `process.pid`) when the HTTP server starts, to see what processes are being created.

Before we create a cluster for this server and fork many workers for it, let's do a simple *load testing* for it. Load testing is basically sending many requests for the server and reporting how much load it was able to handle.

There are many tools you can use to perform load testing, from basic ones like ApacheBench (AB) to fully featured ones like Artillery. To keep things simple for our example, we'll use the simple `loadtest` npm package.

After installing the package, run the HTTP server and then run the `loadtest` command (in a different terminal) to perform the load testing:

```
$ node slow-server.js
```

```
$ npx loadtest http://localhost:3000/
```

Figure 9-1 shows the performance summary I got when I did the testing on my computer.

The single-node server was able to handle about 15 requests per second on my computer. This result will be different on different computers.

```
JS slow-server.js  ×

JS slow-server.js  >  ⊗ createServer() callback

   3    createServer((req, res) => {
   4      for (let i = 0; i < 1e8; i++) {
   5      |   // Simulate CPU work:
   6      }
   7
   8      res.end();
   9    }).listen(3000, () => {
  10      console.log(`Process ${process.pid}`);
  11    });
  12

TERMINAL

$ node slow-server.js          ● $ npx loadtest http://localhost:3000
Process 80442                    Requests: 20, requests per second: 4, mean latency: 1
                                 Requests: 15, requests per second: 3, mean latency: 2
                                 Requests: 13, requests per second: 3, mean latency: 2
                                 Requests: 19, requests per second: 4, mean latency: 2
                                 Requests: 10, requests per second: 2, mean latency: 2

                                 Target URL:          http://localhost:3000
                                 Max time (s):        10
                                 Concurrent clients:  50
                                 Running on cores:    5
                                 Agent:               none

                                 Completed requests:  155
                                 Total errors:        0
                                 Total time:          10.007 s
                                 Mean latency:        2706.8 ms
                                 Effective rps:       15
```

Figure 9-1. Performance summary for the single-node server

Now that we have a reference benchmark for a single node, let's see what difference the node:cluster module will make for this performance.

Here's the code (added to a *cluster.js* file) we need to create a primary process with the node:cluster module and make that process fork many worker processes to run the same HTTP server:

```
import cluster from 'node:cluster';
import os from 'node:os';

if (cluster.isPrimary) {
  const cpus = os.availableParallelism();

  console.log(`Forking for ${cpus} CPUs`);
  for (let i = 0; i < cpus; i++) {
    cluster.fork();
```

```
    }
  } else {
    import('./slow-server.js');
  }
```

We first imported both the node:cluster module and the node:os module. We used the node:os module to figure out how many parallel processes we can run using the availableParallelism function. Depending on the server where the code is run, this function will return a number between one and the maximum number of CPU cores available on that server (os.cpus().length). The plan is to fork one worker process for each available CPU core.

The node:cluster module gives us the handy Boolean flag isPrimary to determine if this *cluster.js* file is being loaded as a primary process. The first time we execute this file, we will be executing the primary process, and that isPrimary flag will be set to true. In this case, we can instruct the primary process to fork our server as many times as we have CPU cores using the cluster.fork method with a loop over the number of CPUs the server has. This makes the cluster take advantage of all the available processing power.

When the cluster.fork line is executed from the primary process, the current file, *cluster.js*, is run again, but this time in *worker mode* with the isPrimary flag set to false.

There is actually another flag set to true in this case (the isWorker flag) if you need to use it.

When the application runs as a worker, it can start doing the actual work. This is where we need to define our server logic, which, for this example, we can do by importing the *slow-server.js* file that we have already.

That's basically it. That's how easy it is to take advantage of all the processing power in a server. To test the cluster, run the *cluster.js* file:

```
$ node cluster.js
Forking for 10 CPUs
Process 15601
Process 15602
Process 15606
Process 15607
Process 15604
Process 15605
Process 15603
Process 15609
```

```
Process 15610
Process 15608
```

The computer I tested with had 10 cores, so the for loop forked 10 worker processes. It's important to understand that these forks are completely different Node processes. Each forked worker process will have its own memory space and event loop.

When this server gets multiple requests, thanks to the built-in load balancing of the node:cluster module, the requests will be handled by different worker processes with different process IDs. The workers will not be exactly rotated in sequence because the node:cluster module performs some optimizations when picking the next worker, but the load will be somehow distributed among the different worker processes.

Now, let's use the same loadtest command to measure the performance of this cluster, as shown in Figure 9-2.

```
JS slow-server.js  ✕                  ···        JS cluster.js  ✕

JS slow-server.js  >  ⊕ listen() callback         JS cluster.js  > ...

 3    createServer((req, res) => {         4    if (cluster.isPrimary) {
 4      for (let i = 0; i < 1e8; i++) {    5      const cpus = os.availablePar
 5        // Simulate CPU work:            6
 6      }                                  7      console.log(`Forking for ${c
 7                          —              8      for (let i = 0; i < cpus; i+
 8      res.end();                         9        cluster.fork();
 9    }).listen(3000, () => {             10      }
10      console.log(`Process ${process.pi 11    } else {
11    });                                 12      import('./slow-server.js');
12                                         13    }

TERMINAL

○ $ node cluster.js            Requests: 130, requests per second: 26, mean latency: 372
  Forking for 10 CPUs          Requests: 130, requests per second: 26, mean latency: 366
  Process 87097                Requests: 126, requests per second: 25, mean latency: 383
  Process 87090                Requests: 126, requests per second: 25, mean latency: 382
  Process 87099                Requests: 120, requests per second: 24, mean latency: 389
  Process 87095                Requests: 259, requests per second: 26, mean latency: 391
  Process 87092
  Process 87093                Target URL:          http://localhost:3000
  Process 87094                Max time (s):        10
  Process 87091          ▮     Concurrent clients:  50
  Process 87096                Running on cores:    5
  Process 87098                Agent:               none

                               Completed requests:  1270
                               Total errors:        0
                               Total time:          10.017 s
                               Mean latency:        385.3 ms
                               Effective rps:       127
```

Figure 9-2. Performance summary for the server cluster

The same computer that handled only 16 requests per second with a single process is now able to handle 123 requests per second! This is a more than sevenfold increase in performance, and all we had to do to make that happen was add a few simple lines of code.

> We didn't have to run the worker processes on different ports. The `node:cluster` module takes care of making the worker processes share ports.

Broadcasting Messages

Communicating between the primary process and the workers is simple because under the hood the `node:cluster` module is using the `child_process.fork` API, which means we also have IPC channels available between the primary process and each worker process.

We can access the list of worker objects using `cluster.workers`, which is an object that holds a reference to all workers and can be used to read information about these workers. Since the communication channels between the primary process and all workers are already established, to broadcast a message to all workers we can use a simple loop over them and use the `send` method on each:

```
Object.values(cluster.workers).forEach(worker => {
  worker.send(`Hello Worker ${worker.id}`);
});
```

The `Object.values` method is used here to get an array of all workers from the `cluster.workers` object. Then, for each worker, we use the `send` function to send over any value we want. In a worker file—`slow-server.js`, in our example—to read a message received from this primary process, we can register a handler for the `message` event on the `process` object. Here's an example:

```
process.on('message', msg => {
  console.log(`Message from primary: ${msg}`);
});
```

Figure 9-3 shows what I see when I test these two additions to the *cluster.js/slow-server.js* example. I added a `setTimeout` around the sending to make sure all the forking is done before sending.

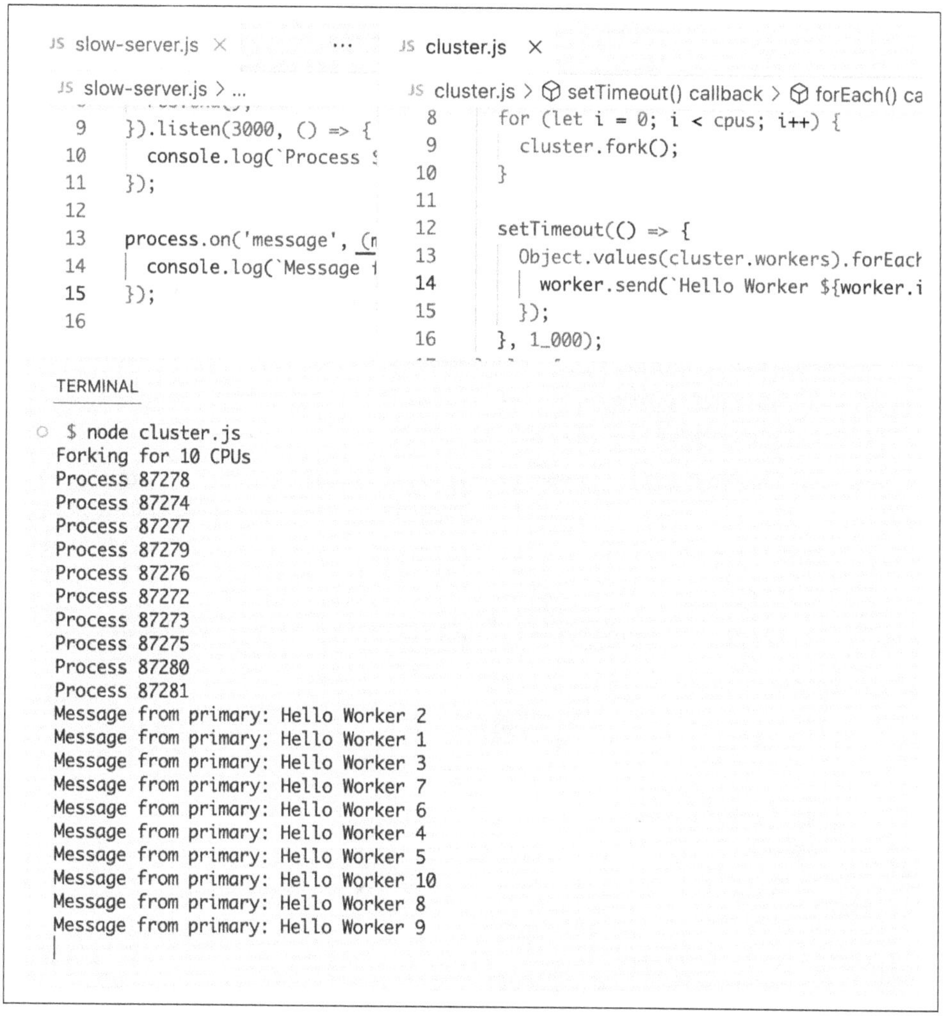

```
JS slow-server.js ×              ···   JS cluster.js ×

JS slow-server.js > ...                JS cluster.js > ⊕ setTimeout() callback > ⊕ forEach() ca
   9    }).listen(3000, () => {         8    for (let i = 0; i < cpus; i++) {
  10      console.log(`Process ?        9      cluster.fork();
  11    });                            10    }
  12                                    11
  13    process.on('message', (n       12    setTimeout(() => {
  14    | console.log(`Message 1        13      Object.values(cluster.workers).forEach
  15    });                            14      | worker.send(`Hello Worker ${worker.i
  16                                    15    });
                                        16  }, 1_000);

TERMINAL

○ $ node cluster.js
  Forking for 10 CPUs
  Process 87278
  Process 87274
  Process 87277
  Process 87279
  Process 87276
  Process 87272
  Process 87273
  Process 87275
  Process 87280
  Process 87281
  Message from primary: Hello Worker 2
  Message from primary: Hello Worker 1
  Message from primary: Hello Worker 3
  Message from primary: Hello Worker 7
  Message from primary: Hello Worker 6
  Message from primary: Hello Worker 4
  Message from primary: Hello Worker 5
  Message from primary: Hello Worker 10
  Message from primary: Hello Worker 8
  Message from primary: Hello Worker 9
```

Figure 9-3. Messages between primary and worker processes

Every worker process received a message from the primary process.

Let's make this communication example a bit more practical. Let's say we need our HTTP service to reply with the number of users we have in the database. To keep things simple, we can use a fake function to replace the database user count logic:

```
// Fake double
const dbUsersCount = (() => {
  let count = 1;
  return () => {
    count = 2 * count;
    return count;
```

```
  };
})();
```

This function simply returns 2 the first time it's called, then doubles its returned value for each subsequent call.

This is just for simplified testing, but since the real scenario here is that the function will make a database request, we need to avoid making multiple requests by the multiple workers we have. We can use some sort of caching in the worker processes and update the cache after a certain period of time (say 60 seconds). This is better, but the 10 workers would still be doing 10 database requests every 60 seconds.

Wouldn't it be great if another process did the database requests and shared the count with all workers every 60 seconds? This is where we start combining the cloning strategy with the decomposing strategy!

For simplicity, let's make the primary process carry out the request to count users in the database (ideally, another entity should do that). We'll make the call to dbUser sCount in *cluster.js* and then use a for loop to broadcast the count value to all workers every 60 seconds.

Here's how I changed the *cluster.js* file to implement that:

```
// In the cluster.isPrimary branch:

const dbUsersCount = (() => {
  let count = 1;
  return () => {
    count = 2 * count;
    return count;
  };
})();

const updateWorkers = () => {
  const usersCount = dbUsersCount();
  Object.values(cluster.workers).forEach((worker) => {
    worker.send({ usersCount });
  });
};

updateWorkers();
setInterval(updateWorkers, 60_000);
```

Here we're invoking updateWorkers for the first time and then invoking it every 60 seconds using a setInterval. This way, every 60 seconds, all workers will receive the new user count value over the IPC channel, and only one database connection will be made every 60 seconds.

In the server code, we can use the `usersCount` value in the `message` event handler. We can simply store that value in a top-scope variable and use it anywhere in that module.

Here's how I changed *slow-server.js* to implement that:

```
let usersCount;

createServer((req, res) => {
  for (let i = 0; i < 1e8; i++) {
    // Simulate CPU work:
  };

  res.write(`Process ${process.pid}\n`);
  res.end(`Users: ${usersCount}`);
}).listen(3000, () => {
  console.log(`Process ${process.pid}`);
});

process.on('message', (msg) => {
  usersCount = msg.usersCount;
});
```

I changed the server response to include the `usersCount` value. If you test the cluster code now, during the first 60 seconds you'll get 2 as the users count from all workers (and only one database request will be made). After 60 seconds, all workers would start reporting the new user count, 4 (and only one other database request would be made).

This is all possible thanks to the communication channels between the primary process and all workers.

Increasing Availability

One of the problems in running a single instance of a Node application is that when that instance crashes, it has to be restarted. This causes some server downtime between these two actions, even if the process was monitored and a restart is automated.

This also applies to the case when the server has to be restarted to deploy new code. With one instance, there will be server downtime that affects the availability of the system.

When we have multiple instances, the availability of the system can be easily increased with only a few extra lines of code. Let's make that happen.

First, to make the example practical for testing, let's simulate a random crash in the server process. We can do a `process.exit` call inside a timer that fires after a random amount of time:

```
// In slow-server.js

setTimeout(() => {
  process.exit(1);
}, Math.random() * 10_000);
```

When a worker process exits like this, the primary process will be notified using the `exit` event on the `node:cluster` object. We can register a handler for that event and simply fork a new worker process to replace the crashed process:

```
// In the cluster.isPrimary branch:

cluster.on('exit', (worker) => {
  if (worker.exitedAfterDisconnect === true) return;

  console.log(
    `Worker ${worker.id} crashed.` +
      'Starting a new worker...',
  );

  cluster.fork();
});
```

Note how I added an `if` condition to make sure the worker process actually crashed and wasn't manually disconnected and killed by the primary process itself. For example, the primary process might decide that we are using too many resources based on the load patterns it sees, and it will need to kill a few workers in that case. To do so, we can use the `disconnect` methods on any `worker` object and, in that case, the `exitedAfterDisconnect` flag will be set to `true`, so the `if` statement will guard against forking a new worker for that case.

If we run *cluster.js* with the preceding handler (and the random crash in *slow-server.js*), after a random number of seconds, workers will start to crash, and the primary process will immediately fork new workers to increase the availability of the system. You can actually measure the availability using the same `loadtest` command and see how many requests the server will not be able to handle overall (because some of the unlucky requests will still have to face the crash case).

Figure 9-4 shows the output of `loadtest` on my computer while the random crash was happening.

```
JS slow-server.js ×          □ ...        JS cluster.js  ×

JS slow-server.js > ...                   JS cluster.js > ...
  0      }                          12        cluster.on('exit', (worker) => {
  7                                  13          if (worker.exitedAfterDisconnec
  8        res.end();                14
  9      }).listen(3000, () => {     15          console.log(
 10        console.log(`Process ${proces 16           `Worker ${worker.id} crashed.
 11      });                         17            'Starting a new worker...',
 12                                   18          );
 13      setTimeout(() => {     ─    19
 14        process.exit(1);          20          cluster.fork();
 15      }, Math.random() * 10_000); 21        });
 16                                   22      } else {
                                      23        import('./slow-server.js');
                                      24      }

TERMINAL

Worker 9 crashed.Starting a new worker...      $ npx loadtest http://localhost:3
Worker 4 crashed.Starting a new worker...
Process 88344                               Target URL:          http://localh
Worker 8 crashed.Starting a new worker...   Max time (s):        10
Process 88347                               Concurrent clients:  50
Process 88348                               Running on cores:    5
Worker 12 crashed.Starting a new worker...  Agent:               none
Process 88350
Worker 15 crashed.Starting a new worker...  Completed requests:  1228
Process 88354                               Total errors:        16
Worker 10 crashed.Starting a new worker...  Total time:          10.003 s
Process 88355                               Mean latency:        327.7 ms
Worker 17 crashed.Starting a new worker...  Effective rps:       123
Process 88356
```

Figure 9-4. Performance summary while the server is randomly crashing

Only 16 requests failed out of 1,228. That's over 98% availability. By adding a few simple lines of code, we now don't have to worry about process crashes anymore. The primary process will keep an eye on the failed processes for us and replace them on demand.

Zero-Downtime Restarts

When we deploy new code to production servers, all Node processes need to be restarted. During a restart, requests to these processes will fail.

In a cluster, instead of restarting all worker processes together, we can restart them one at a time. This way, while one worker is being restarted, the others will continue to serve requests, and we won't have any server downtime at all.

Implementing this with the `node:cluster` module is actually really easy. Let me show you how.

First, since we don't want to restart the primary process once it's up (because that would be total downtime for all workers), we need a way to send the primary process a command to instruct it to start restarting its workers (one by one).

There are a few ways to do that depending on what OS the server is running. With Linux systems, we can listen to a process signal like `SIGUSR1` or `SIGUSR2`. These signals can be sent to a running process by issuing a `kill` command like this:

```
$ kill -SIGUSR2 PID
```

In Node, we can listen to this `SIGUSR2` signal as an event on the `process` object, and within its handler we can do whatever we want:

```
process.on('SIGUSR2', () => {
  // Do something
});
```

> Don't use `SIGUSR1`. Node uses that for debugging purposes. Also, on Windows, these process signals are not supported, and you would have to find another way to instruct the primary process to do something. There are some alternatives. For example, you can use standard input or socket input. You can also monitor the existence of a `process.pid` file and track any remove events on it.

Now that we know how to instruct the primary process to restart its workers, we can have it do so by disconnecting each worker and forking a new one to replace it. However, since we want it to disconnect one worker at a time, we need to have it disconnect one worker, wait for that, fork another worker, wait for the fork to start taking requests, then disconnect the next worker.

A worker process has an `exit` event similar to the `node:cluster` event but specific to the worker, and it means that specific worker has exited. We can use that `exit` event to start the forking of the next worker.

A worker process also has a `listening` event that is triggered after a call to `listen` happens in that worker. We can use that event to start disconnecting the next worker.

Putting it all together, here are the changes I made to *cluster.js* to implement the gradual reloading of all workers:

```
// In the cluster.isPrimary branch:

console.log(
  `To restart workers, use: kill -SIGUSR2 ${process.pid}`,
);
```

```
process.on('SIGUSR2', () => {
  const workers = Object.values(cluster.workers);

  const restartWorker = (workerIndex) => {
    const worker = workers[workerIndex];
    if (!worker) return;

    worker.on('exit', () => {
      if (worker.exitedAfterDisconnect === false) return;

      console.log(`Exited process ${worker.process.pid}`);

      cluster.fork().on('listening', () => {
        restartWorker(workerIndex + 1);
      });
    });

    worker.disconnect();
  };

  restartWorker(0);
});
```

The workers array has the list of all cluster workers. The restartWorker function receives the index of a worker to be restarted, starting from 0. This way, we can do the restarting in sequence by having this function call itself with the next index value when it's ready for the next worker.

In the restartWorker function, we got a reference to the worker to be restarted. Since we will be calling this function recursively to form a sequence, we need a stop condition. When we no longer have a worker to restart, we can simply return. We then disconnect that worker (using worker.disconnect), but before restarting the next worker we need to fork a new one to replace this current disconnecting one. We can use the exit event on the current worker to fork a new worker when the current one exists.

We have to make sure that the exit event was actually triggered after a normal disconnect call (the opposite of the crash example). We can use the exitedAfter Disconnect flag. If this flag is false, the exit was caused by something other than a disconnect call, and in that case, we should return and do nothing. But if the flag is true, we can go ahead and fork a new worker to replace the one that we're disconnecting.

When the new forked worker is ready, which is determined here using the listening event, we can restart the next one by calling the restartWorker function again with the next index value for the workers array.

That's all we need for a zero-downtime restart. To test it, I added a `console.log` statement to give us the ID of the cluster process.

Running the logged `kill -SIGUSR2 PID` command should start the workers' reloading process.

To see the zero-downtime in action, before running the kill command, run the same `loadtest` command. Figure 9-5 shows the results when I tested this.

```
JS slow-server.js ×                  □  ···     JS cluster.js  ×

JS slow-server.js > ...                         JS cluster.js > ...
 1   import { createServer } from '|        8      for (let i = 0; i < cpus; i++) {
 2                                          9          cluster.fork();
 3   createServer((req, res) => {          10      }
 4     for (let i = 0; i < 1e8; i++        11
 5       // Simulate CPU work:             12      console.log(
 6     }                                   13          `To restart workers, use: kill
 7                                         14      );
 8     res.end();                         15
 9   }).listen(3000, () => {              16      process.on('SIGUSR2', () => {
10     console.log(`Process ${proces      17          const workers = Object.values(c
11   });                                  18
12                                         19          const restartWorker = (workerIn
                                           20            const worker = workers[worker

TERMINAL

○ $ node cluster.js                          ● $ npx loadtest http://localhost:3|
  Forking for 10 CPUs                          Requests: 121, requests per second
  To restart workers, use: kill -SIGUSR2 88792 Requests: 126, requests per second
  Process 88794                                Requests: 122, requests per second
  Process 88797                                Requests: 120, requests per second
  Process 88795                                Requests: 122, requests per second
  Process 88796                                Requests: 243, requests per second
  Process 88800
  Process 88793                                Target URL:          http://localh
  Process 88799                                Max time (s):        10
  Process 88798                                Concurrent clients:  50
  Process 88801                                Running on cores:    5
  Process 88802                                Agent:               none
  Exited process 88793
  Process 88839                                Completed requests:  1235
  Exited process 88794                         Total errors:        0
  Process 88840                                Total time:          10.004 s
  Exited process 88795                         Mean latency:        395.6 ms
  Process 88843                                Effective rps:       123
  Exited process 88796
```

Figure 9-5. Performance summary while the workers are restarted

No requests failed at all during the reloading of this cluster!

Handling State

One of the important challenges to understand in a scaled environment is how to handle in-memory state and any stateful communications in general.

Since the workers have their own separate memory spaces, we cannot cache things in memory in one worker. The other workers will not have access to that cached value.

If you need to cache things in a scaled environment, you have to use a separate entity and read/write to that entity's API from all workers. This entity can be a database server, a service like Redis, or a dedicated Node process with a read/write API for all other workers to communicate with.

This is known as *distributed caching*, and it's an example of the decomposing scalability strategy for an application. This is something that you should do even if you're running your application on a single core.

Another example of state management is handling user authentication and managing user sessions. In a scaled environment, a request for authentication comes to the primary process and gets sent to one worker. That worker will start recognizing the state of the user. However, when the same user makes another request, the load balancer will eventually send them to other workers, which do not have them as authenticated. Keeping a reference to an authenticated user session in one instance is not going to work anymore. Instead, we need to keep workers stateless and manage stateful information in an external entity.

Another way to deal with the user sessions problem is to use what's known as *sticky sessions*. It's not as efficient as using an external stateful entity, but it's certainly easier to implement. We can simply use a routing algorithm like IP hash.

Sticky sessions is a simple example of the splitting scaling strategy. When a user authenticates with a worker process, a record of that relation is kept in the primary process. Then, when the same user sends a new request, we use that record to keep sending them to the same worker process. This way, the code on the server side does not have to be changed.

While a routing algorithm like IP hash is not natively supported by Node's `node:cluster` module, implementing it is fairly simple. We can put the workers in an array, reduce the IP of an incoming request into an index within that array, and have the worker with that index handle the request. Here's a very basic implementation of that:

```
import os from 'node:os';
import { fork } from 'node:child_process';
import { createServer } from 'node:http';

const cpus = os.availableParallelism();
const workers = [];
```

```
for (let i = 0; i < cpus; i++) {
  workers.push(fork('./slow-server.js'));
}

function getWorkerIndex(ip) {
  const hash = ip.split('.').reduce(
    // Hash IP octets and combine them
    (hash, part) => Number(part) + 256 * hash,
  );
  return hash % cpus;
}

createServer((req, res) => {
  const ip = req.socket.remoteAddress;
  const workerIndex = getWorkerIndex(ip);
  workers[workerIndex].send({ req, res });
  // IPC logic
}).listen(3000);
```

While using a sticky load-balancing algorithm like IP hash is simple to implement and can be suitable and efficient for some cases (like a small application with a few workers), it's certainly not ideal. It can lead to uneven workload distribution, and it doesn't scale very well. Keeping Node workers stateless and managing state with an external service is a better approach in terms of workload distribution and scalability.

Sticky load-balancing algorithms like IP hash can be easily configured and used in other load balancers, like nginx, and many others.

Process Managers

If you don't want to manage your own cluster code, you can use one of the advanced process managers packages available for Node. These packages wrap the node:cluster module and provide a CLI to simplify the management of forked processes.

One of the popular advanced process managers for Node is PM2, and while most of its features cost money to use, it offers the basics of running and managing a cluster of workers for free.

Install the pm2 package with npm and take a look at its help page (with a -h option). It has many commands and options to start, monitor, and manage a cluster (along with many other features).

For example, to launch a cluster of workers for a Node process, you just start the Node process with the start command, and use the -i option (with either a custom number of forks, or max to use the maximum possible forks):

```
$ npx pm2 start script.js -i max
```

To perform a zero-downtime restart, you can use the `reload` command:

```
$ npx pm2 reload all
```

To see a list of all worker processes with their status and usage, you can use the `list` command:

```
$ npx pm2 list
```

Summary

Scaling in Node is a first-class concept. It's why Node is named Node, after all.

There are many scaling strategies like cloning, decomposing, and splitting. The focus of this chapter was mainly on the cloning strategy.

Node's built-in `node:cluster` module can be used to create a primary process that forks many worker processes and manages them. It can start and stop them, send them data, and receive data from them. Workers can share ports, and the primary process can distribute the workload among them to increase the performance of the server.

The workload can be distributed among worker processes in multiple ways. The `node:cluster` module uses a simple cyclical routing algorithm to load balance incoming requests among workers, but other routing algorithms can be used. Caching and managing stateful communication in general can be done by using other routing algorithms or by using external services.

The primary process can also monitor the workers, replace faulty ones immediately, and restart workers one by one. This improves the uptime and fault tolerance of the server.

In the next chapter, we'll explore some of the tools that can be used to improve development workflows and help with building, testing, deploying, and maintaining Node code.

CHAPTER 10

Practical Node

While Node's core modules are designed to help us create backend services, their flexibility, asynchronous APIs, and easy integration with external environments make Node a great environment to run tools that help with development workflows on both the backend and the frontend. In this chapter, we'll explore some of these tools and understand their roles in building, testing, deploying, and maintaining Node projects.

These tools range from small libraries that focus on a few tasks to big frameworks that can be used for full stack development. For each specific task, there are many tools that you can use. In this chapter, I'll focus on the most popular tools and provide an overview of their value and the basics of how to use them.

We'll talk about code quality tools, module bundlers, task runners, web frameworks, and JavaScript extensions and transpilers.

Code Quality Tools

In Chapter 1, we reviewed a few Node tools that can be used to assist in the lifecycles of an application. Let's expand on two of the most important ones when it comes to code quality, Prettier and ESLint.

Because of the great value these tools add to a project, I would argue that no Node project should be built without them. They are simple and easy to integrate into a Node project, and with them you make faster progress and write cleaner code with consistent standards. Why would anyone not use something that makes them faster and better?

Prettier is used to automate the formatting of code. No matter how good one is with manually formatting code, without Prettier, the formatting will be inconsistent. More importantly, why spend time manually formatting the code anyway? And why spend time debating how the formatting should be in code reviews?

Prettier can format code in many languages, not just JavaScript. For frontend projects, you can also use it to format CSS, HTML, JSON, GraphQL, and even Markdown.

ESLint can be used to enforce code quality rules and find any potential errors in your code as early as when you type it. It's like a detective that's always with you, looking over your shoulders and nudging you every time you try to do something wrong.

ESLint is highly configurable. You can make it super strict or super flexible. You can pick your own rules or use the recommended ones. You can make it autofix some problems or just report them as you type. You can tell it to ignore certain violations in general or for specific reasons.

There are many sets of recommended rules organized by different developer communities. I like to use the built-in ones. You can see a list of all ESLint rules and which ones are recommended on the ESLint website (*https://eslint.org/docs/rules*).

> You might not understand some of these recommended rules. Look at that as an opportunity to learn something new! These rules exist for good reasons. Look them up and learn why they exist.

The best way to integrate Prettier and ESLint into your workflow is to make them run every time you save your code. Some even prefer to have them run while they're typing code! Some like to run them before pushing changes. I like the first option. Many code editors have plugins and settings to make these tools run on save or while you type.

Having Prettier and ESLint integrated into code editors is great, but the project should also automatically enforce the use of its tools. Maybe an editor plugin is malfunctioning. Maybe someone on the team forgot to activate it or accidentally saved while the plugin was disabled.

Enforcement can happen locally with hooks before the code is pushed upstream, or it can happen remotely after the code is pushed but before it gets accepted. We'll explore some tools that can be used for enforcement later in this chapter.

The npm `prettier` package installs a CLI that can be run to format the code and to check if the code is formatted correctly. The same is true for the `eslint` package.

These two packages overlap a bit in their offerings. For example, you can enforce the use (or nonuse) of semicolons in JavaScript using both, but these overlaps are just a small part of their features and nothing to worry about. ESLint and Prettier work great together for both backend and frontend code, and their packages should be among the first you add in your development environment.

> There are a few arguments against using Prettier and ESLint. I understand that they can be annoying sometimes. They are not perfect all the time. They might force you to rewrite code that is fine the way it is. Basically, sometimes they will make you spend more time than necessary, but in my opinion, the value they bring to the table most of the time trumps the little imperfections they sometimes have. They are also getting better over time, and you can customize them to be more flexible when you need to.

Prettier

To use Prettier in a Node project, you can install it with npm:

```
$ npm i -D -E prettier
```

Note how it is a development dependency. It plays no role in running the application in production.

Note also the `-E` flag to save this dependency as an exact version. This ensures that everyone working on the project uses the same version of Prettier. Even patch releases of Prettier can result in slightly different formatting. Updating Prettier should be done periodically in an isolated change specific to that task.

Prettier has a set of default configurations and can be used out of the box, but if you want to customize the way it works, you can configure it for a Node project using one of the following ways:

- A `prettier` key in *package.json*
- A JavaScript module file with the name *prettier.config.js*
- A JSON configuration file with the name *.prettierrc*

There are more ways to configure Prettier, but these are the most common.

Here's an example Prettier JSON configuration file:

```
{
  "arrowParens": "always",
  "quoteProps": "consistent",
  "singleQuote": true,
  "trailingComma": "es5"
}
```

The `singleQuote` option, for example, formats the code to use single quotes instead of double quotes around strings. The `always` value for `arrowParens` formats the code to always use parentheses for arrow functions even when they are not needed. You can see a complete list of all options and what they mean in the Prettier docs (*https://prettier.io/docs/en*).

To use Prettier, you run the `prettier` command with either a `--write` option to rewrite the files with the new formatting or a `--check` option to check if all the files are formatted correctly. The latter can be used to continuously enforce the use of Prettier in the project.

The `prettier` command can be run on a file, a folder, or all folders within the project using `.` as the target:

```
$ npx prettier -w .
```

ESLint

To use ESLint in a Node project, you need to install it and add a configuration file. You can do both with this command:

```
$ npm init @eslint/config@latest
```

This command will ask a few questions interactively and use your answers to create an *eslint.config.js* file. It'll install ESLint (and other related packages) as development dependencies in your project.

You can customize ESLint globals, plugins, and rules within its configuration file. Here's a configuration file showing examples of each:

```
import globals from 'globals';
import pluginJs from '@eslint/js';

export default [
  {
    languageOptions: {
      globals: { ...globals.browser, ...globals.node },
    },
  },
  pluginJs.configs.recommended,
  {
    rules: {
      'curly': 'error',
      'no-else-return': 'error',
      'no-unneeded-ternary': 'error',
      'no-useless-return': 'error',
      'no-var': 'error',
      'prefer-const': 'error',
      'yoda': ['error', 'never', { exceptRange: true }],
    },
  },
];
```

By default, ESLint will complain if your code is using a variable that's not declared. This configuration makes an exception for the global variables for both Node and browser environments, so using global scope objects like `process`, `console`, and `window` will be permitted.

The recommended rules are added using the `@eslint/js` plugin, then a set of custom rules (which are not part of the recommended ones) are added as well. You can also use a custom rule to override a recommended one.

To mention a few examples of custom ESLint rules: The `curly` rule ensures that all block statements are wrapped in curly braces, even when they're not syntactically needed. The `no-var` rule ensures that the `var` keyword is not used to declare variables. The `yoda` rule blocks any use of yoda conditions in `if` statements. Note how this rule has more configuration to make an exception for range conditions. If any of these custom rules pique your interest, you can look them up in the ESLint rules reference (*https://eslint.org/docs/rules*).

Once ESLint is installed and configured for a project, if your editor has ESLint support it will start showing you any ESLint problems. The `eslint` package also comes with a command that you can run locally (or in a continuous integration pipeline) to report any problems or make sure that there are none:

```
$ npx eslint .
```

Other Tools

Prettier and ESLint are great, and they will improve the quality of your code, but there are many other tools that you can use in Node projects for code quality as well:

Testing frameworks
> While Node has great built-in fundamentals for testing, there are many great Node testing libraries that offer many more features and easier-to-use APIs. Mocha and Chai are among the most popular, but my favorite is Jest. Jest is simple and comprehensive. It's great for both backend and frontend projects.

Static typing tools
> While JavaScript dynamic types make it feel more flexible, in reality the lack of static types (that can be checked at compile time) opens many paths to hidden problems in your code. Many tools evolved to add static typing to JavaScript. The most popular and featured one is TypeScript.

Editors and AI assistants
> An advanced code editor makes a big difference in the quality of your code. Editors like WebStorm, Atom, and Visual Studio Code (VS Code) offer great code quality features like intelligent code completion and navigation, error detection, integrated debugging and source control, and many more. AI code assistants like

GitHub Copilot make these features even more powerful. They leverage machine learning from open source code to provide context-aware code suggestions for best practices and coding standards.

Module Bundlers

For a frontend environment like a web browser, requesting a resource (like a JavaScript or CSS file) from a server is an asynchronous task done over the network. Every network request affects the performance of the web browser and the quality of the user experience. This is especially true on mobile devices and devices with limited resources and connectivity in general.

> While developing frontend applications, you should consider testing them with limited resources and connectivity. Browsers like Chrome offer a throttling mode to restrict both the network and CPU. These restrictions make you aware of how your application will behave on limited resources and connectivity. This helps you make better decisions and handle cases you would not be aware of otherwise.

While you can make your modules available individually and things would work fine in modern browsers (which support importing modules), you should still try to minimize the number (and size) of resources needed to run your application on frontend devices.

The following are some of the common practices that can help with minimizing the number and size of application resources:

- Combine all the initial resources needed to run the first view of the application and serve them all with one request. This is known as *bundling*, and the one file generated is known as a *bundle file*.
- Minify the code in the bundle file by removing whitespaces and shortening variable names to reduce its size. While your source code should be as readable as possible, the code you ship to browsers does not need to be.
- Remove any unused code. This could be part of your own code, like a function you defined but never used, or a library you installed but used only a few of its features. For example, you might install lodash in an application and use only a handful of its functions. You don't need the rest of its functions in the bundle. All code that's not used in your application can be removed to reduce the bundle size. This process is known as *tree shaking*.
- After the first view, other tasks in your application can request more bundles based on what they do. Instead of shipping the entire code of your application in

a single bundle, you can split your code by feature into different bundles, and load these bundles on demand when they are needed. This is known as *code splitting*.

There are multiple Node tools (such as Webpack, Parcel, and Rollup) that can help with all of these practices (and more). Let's explore the basic use of Webpack for module bundling.

Webpack is split into two main packages, one for the core library (`webpack`), and one for the CLI (`webpack-cli`). Install both of them as development dependencies:

```
$ npm i -D webpack webpack-cli
```

To run Webpack, you need to give it an *entry point*. This is the first file that Webpack will start with. It's the start of your dependency graph.

> You can actually run the `webpack` command without any configuration if your application entry point is *src/index.js*.

An application can also have multiple entry points, and Webpack can generate multiple bundles. You can use Webpack with loaders to make it process many file types before bundling. It can, for example, process CSS files (and CSS extensions). It can process Babel and TypeScript. It can even process things like SVG, JSX, GraphQL, and many more.

Webpack is built with a plugin-friendly structure, and it supports the use of external plugins that can hook into Webpack's lifecycle to implement a variety of features like defining global constants, copying files, ignoring modules, and even integrating with tools like ESLint and Prettier.

Here's a simplified example Webpack configuration file for a project that uses Type-Script and the Sass CSS extension:

```
const webpack = require('webpack');

const CopyPlugin = require('copy-webpack-plugin');
const MiniCssExtractPlugin = require('mini-css-extract-plugin');

module.exports = {
  entry: {
    app: './src/app.ts',
    search: './src/search.ts',
  },
  output: {
    filename: '[name].ts',
    path: __dirname + '/dist',
```

```
    },
    module: {
      rules: [
        {
          test: /\.ts$/,
          exclude: /node_modules/,
          use: { loader: 'ts-loader' },
        },
        {
          test: /\.scss$/,
          exclude: /node_modules/,
          use: [
            MiniCssExtractPlugin.loader,
            { loader: 'css-loader', options: { url: false } },
            'sass-loader',
          ],
        },
      ],
    },
    plugins: [
      new CopyPlugin({
        patterns: [{ from: 'src', to: 'dist' }],
      }),
      new webpack.DefinePlugin({
        'PRODUCTION': JSON.stringify(process.env.NODE_ENV === 'production'),
        'process.env.API_KEY': JSON.stringify(process.env.API_KEY),
      }),
    ],
  };
```

This configuration instructs Webpack to use two entry point files and produce two bundle files matching their names. It uses a TypeScript loader to process all files ending in .ts, and a Sass loader to process all files ending in .scss. It also extracts CSS files into separate bundles using a loader from the mini-css-extract-plugin package. Finally, it uses the copy-webpack-plugin package to copy static files from src into dist, uses the built-in DefinePlugin to define a PRODUCTION flag, and enables the use of process.env.API_KEY environment variable in the bundle.

As you can see, Webpack is very flexible and has many options and features to customize how it works. Other module bundlers offer similar (and more) features and differentiate themselves in performance and the size of the bundle they generate as well.

> The HTTP/2 protocol, with its multiplexing, compression, and server push features, reduces the need for bundling to some extent since browsers can handle multiple simultaneous requests and can cache them better individually. Bundling still has its benefits, but as browsers evolve, this will be a balancing act between what to bundle and what to ship individually.

Task Runners

Task runner tools like gulp and Grunt are used to run (and automate) repetitive tasks in different environments. Some tasks are needed in development environments, some are needed in production, and some are used to deploy changes in code to different environments and servers.

Here are some examples of tasks that need to be done regularly:

- Formatting and checking code for problems
- Running tests and analyzing coverage
- Minifying code and generating bundles
- Deploying code to other environments
- Generating reports and sending emails

Doing these tasks manually is time-consuming, and mistakes might be made. Defining these tasks and having a standard way to run them every time reduces the chance of manual mistakes, and it saves time in the process. Task runners can also improve the performance of their defined tasks. For example, gulp uses Node streams under the hood, which makes it perform better, especially for big tasks and those that need to work with the filesystem.

A task defined with a task runner also has the benefit of clarity when someone needs to understand what exactly that task does, or to debug a problem with that task.

Task runners like gulp and Grunt have plugins that can be used to define common tasks. For example, say that an application uses the Sass CSS extension and you need to define a task to transform all Sass files into CSS files. While you can do that with a featured module bundler like Webpack, let's assume that you want to do it differently for your production environment.

You can use the following gulp task:

```
const gulp = require('gulp');
const sass = require('gulp-sass')(require('sass'));

gulp.task('sass', () => {
  return gulp.src('src/*.scss') // Source files
    .pipe(sass())               // Compile Sass to CSS
    .pipe(gulp.dest('dist'));   // Output destination
});
```

Or you can use the following Grunt task:

```
module.exports = function(grunt) {
  grunt.initConfig({
    sass: {
```

```
        dist: {
          files: {
            'dist/app.css': 'src/app.scss'
          }
        }
      }
    });

    grunt.loadNpmTasks('grunt-sass');
    grunt.registerTask('default', ['sass']);
  };
```

For simple tasks, you can use the npm `scripts` section in *package.json*:

```
"scripts": {
  "sass": "sass src/app.scss dist/app.css"
}
```

You can also define and run your tasks in the cloud with continuous integration and deployment services like GitHub Actions, Travis CI, CircleCI, and similar services. These services offer more featured and integrated workflows that can be run concurrently and are highly configurable.

Picking a local task runner or a cloud service is a matter of preference, but regardless of your choice, what's important is that your application tasks should be well-defined, easy to run with simple commands, and automated for most cases.

Automation in Node

You need to maximize the value you get from all the quality and efficiency tools you use for Node, and for that, you need to automate your workflows.

Automation is your guarantee that nothing will get past the guards. Don't rely on the human brain to remember. Automation is also your biggest time-saver. Don't run commands from tools manually; make them autorun at certain points in your workflows.

Integrate the tools with your editor to make them work as you type or when you save changes. Use watch mode when you work with module bundlers and testing frameworks. Define remote pipelines for continuous integration and deployment. Don't go overboard though; the line between productivity and overengineering is a thin one. Beyond the obvious automations for things you know for sure you'll be doing often, wait until you notice that you're repeating a task over and over, then automate it.

There are a few tools that you can use to improve and automate your local workflows in Node. Here are a few examples:

npm pre and post scripts
    Let you run a script before or after another script is executed. For example, a `pretest` script can be used to run ESLint before you run your tests.

---

npm-run-all
> Lets you run multiple npm scripts in parallel or sequentially. You can use it, for example, to run both Prettier and ESLint together.

Husky
> Lets you autorun commands at different stages of the Git lifecycle. You can use it, for example, to run a gulp task every time you commit changes to Git or every time you push commits upstream.

Live Server
> Monitors changes in a web server's project files and reloads the browser automatically.

Frameworks

When it comes to web development, while Node provides excellent built-in support for HTTP through its `node:http` (and `http2/https`) modules, most developers prefer to use higher-level frameworks to build web servers.

Node's built-in support provides only the fundamental building blocks for building and running web servers. Using the built-in support on its own (while possible) is not ideal. You'll need to handle many things manually, and this might result in unstructured, repetitive code that's hard to understand and maintain.

Web frameworks like Express, Koa, hapi, AdonisJS, and many others offer easier-to-use APIs. They offer greater abstractions of common tasks to allow you to focus on your own logic. They are well-structured, battle-tested, and efficient. More importantly, the people who use them keep improving them and adding more plugins/extensions to expand and improve on their functionalities.

One of the common tasks you'll need to handle in a web server is *routing*, which is the process of defining how an application responds to incoming requests for specific URLs (or routes) and HTTP methods. Without a web framework, we can check the `req.url` value and define a switch statement for any route we need to support:

```
import { createServer } from 'node:http';

const server = createServer();

server.on('request', (req) => {
  switch (req.url) {
    case '/':
      // Logic for the main page
      break;
    case '/about':
      // Logic for the about page
      break;
```

```
    // More cases for all app routes

    default:
    // Logic for 404
  }
});

server.listen(3000, () => {
  console.log("Server running at http://localhost:3000/");
});
```

While this might be fine for a few simple pages, and we can improve it for more, it's simply not a good way to implement routing. Just imagine where this will take you when you need to account for HTTP methods as well:

```
server.on('request', (req) => {
  switch (req.url) {
    // ...

    case '/user':
      switch (req.method) {
        case 'GET':
          // Logic for reading user
          break;
        case 'POST':
          // Logic for creating user
          break;
        case 'DELETE':
          // Logic for deleting user
          break;
      }
      break;

    // ...
  }
});
```

You need a higher-level abstraction to simplify a task like this. This is a wheel that was invented long ago by frameworks like Express and others. Here's what routing looks like in Express:

```
import express from 'express';

const server = express();

server.get('/', (req, res) => {
  // Logic for main page
  res.send('Home Page');
});

server.get('/user', (req, res) => {
  // Logic for reading user resource
```

```
    res.send('User Page');
});

server.post('/user', (req, res) => {
  // Logic for creating user resource
  res.send({ status: 'ok' });
});

server.listen(3000, () => {
  console.log('Server running at http://localhost:3000/');
});
```

Run this file after you install Express with npm and test things out. Note a few things about this code:

- The main structure of the web server is similar to the built-in structure. We create a server object and run it using the `listen` method. This is because Express is a wrapper around Node's `node:http` module.
- Instead of having to deal with conditions, we have a more declarative syntax to define different handlers for different URLs and HTTP methods.
- Instead of using low-level methods like `write` and `end` (and others) to prepare and send a response with the right headers, we have a `send` method that takes care of it all. You can even use it to send text, HTML, and JSON objects and arrays. It even works with buffers in Node.

To send static resources like files or images, the many multiline functions that you need with Node's built-in `node:http` module become a single line with Express:

```
// Without Express
fs.readFile(imagePath, (err, data) => {
  if (err) {
    res.writeHead(500, { 'Content-Type': 'text/plain' });
    res.end('Internal Server Error');
    return;
  }
  res.writeHead(200, { 'Content-Type': 'image/png' });
  res.end(data);
});

// With Express
res.sendFile(imagePath);
```

Besides simplifying common tasks, frameworks like Express add many enhanced features to make your code more flexible and easier to maintain. For example, Express has a middleware feature where you can hook any logic into all requests and modify all responses if you need to. Express middleware functions are useful for handling tasks like logging, authentication, parsing, error handling, and more.

Here's an example of a web server using middleware:

```
import express from 'express';
import morgan from 'morgan';

const server = express();

// Built-in middleware for parsing JSON
server.use(express.json());

// 3rd-party middleware for logging
server.use(morgan('tiny'));

// Application-level middleware for authentication
const authenticate = (req, res, next) => {
  const token = req.header('Authorization');
  if (token === process.env.SECRET_TOKEN) {
    next();
  } else {
    res.status(401).send('Unauthorized');
  }
};

server.use(authenticate);

server.get('/', (req, res) => {
  // ...
});
```

Express has many other great features and countless packages that can extend its functionalities. If you are using Node to build a web server, investing time into learning a framework like Express or its alternatives is totally worth it.

You can take things one step further and use a framework that's even more high-level than Express. Many Node web frameworks integrate into the frontend side of web applications. For example, a web application built with React can benefit from having a React Node framework like Next.js, which can be used to implement advanced features like server-side rendering, file-based routing, automatic code splitting, and more.

Many other frameworks have specialized use cases, such as the following:

Apollo Server
    Specialized in building GraphQL-based API servers

Socket.IO
    Specialized in real-time applications

Strapi
    Specialized in content management systems

Mailchimp Open Commerce (formerly Reaction Commerce)
    Specialized in ecommerce applications

NestJS is another important framework for Node that focuses on backend services in general. It has a great modular architecture that makes your code easier to maintain and scale. It uses features like dependency injection and decorators, and it integrates very well with databases like MongoDB and PostgreSQL. NestJS also works great with microservices, and it supports many transport layers (e.g., HTTP, WebSockets, and gRPC).

JavaScript Transpilers

The JavaScript language has been continually improving, especially since the big release of the sixth version of ECMAScript in 2015. Both Node and browser environments have been getting continual updates to support the latest features of the language.

While you, as the developer of a Node application, can easily update and use the latest version of Node and take advantage of the latest features of JavaScript, when you ship your applications to run in browsers, there will be users who use older browsers that do not support modern features.

This is where JavaScript transpilers like Babel and TypeScript can be used to bridge the gap. These transpilers take the JavaScript code that's written with modern syntax (or extended syntax) and convert it into code that's more compatible with a wider range of browsers and their versions.

You can pick how compatible you want this conversion to be. For example, you can configure them to support the most recent three versions of Chrome and the most recent five versions of Microsoft Edge. You would ideally make this decision based on the usage patterns you see for your application.

Besides supporting older browsers, there's also the benefits to consistency and performance. Maybe one browser's implementation of a certain modern feature is slow or incomplete. Transpiling the code ensures that it runs consistently across all browser environments. Transpiling your code can also improve its performance in some cases as the transpilers have many of the best practices built into them and will often write more efficient versions of what your code is trying to do.

Babel focuses on transpiling the modern JavaScript syntax. It also includes polyfills that can be used to add missing functionalities in older environments. Polyfills are great, as they are only used when needed and browsers that do not need them will run the code as is.

The other major JavaScript transpiler is TypeScript, but TypeScript has a much bigger scope than Babel. Besides transpiling modern JavaScript syntax, TypeScript extends the JavaScript language itself to add many useful features like static types, inference, interfaces, decorators, and many more.

TypeScript

In my opinion, the lack of static typing in JavaScript is one of the valid arguments against the language. In JavaScript, types are dynamic. You can define a variable, initialize it with a number, then later make its value a string instead. While one might argue that this makes the language flexible, it often leads to unexpected errors that surface only when the code is run.

Having fixed types for variables and functions (arguments and return values) greatly improves your code quality, readability, and reliability. It makes both debugging and refactoring easier and faster, and it gives developers more confidence to make changes. More importantly, code problems can be detected before the code is run. TypeScript simply makes you more efficient.

When someone starts with TypeScript, they might feel like it is slowing them down. In reality, TypeScript will save you a lot of time the more you use it and the bigger your application gets. I think not using TypeScript in a Node project today is a mistake. Its value is just too great to be missed.

> While there are alternatives to TypeScript, nothing comes close to it in terms of features, integrations, and continuous development.

Let's see a few examples of how TypeScript brings great value to your Node code. First, install it as a development dependency in your project:

```
$ npm i -D typescript @types/node
```

This gives you the `tsc` command. TypeScript is integrated in many editors as well. Once it's installed, an editor like VS Code will start showing you any problems reported by TypeScript. You can also run the `tsc` command to see if there are any TypeScript problems.

Note that we also installed @types/node, which is a type definition package for Node. It defines types for Node's globals and the APIs of its built-in modules. We'll see an example of that shortly.

Next, you'll need a *tsconfig.json* configuration file to customize how TypeScript works. You can use tsc --init to create that file with the recommended settings (and commented-out settings) with their explanations:

```
{
  "compilerOptions": {
    /* Visit https://aka.ms/tsconfig to read more about this file */

    "target": "es2016",
      /* Set the JavaScript language version for emitted JavaScript
        and include compatible library declarations. */

    "module": "commonjs",
      /* Specify what module code is generated. */

    "esModuleInterop": true,
      /* Emit additional JavaScript to ease support for importing
      CommonJS modules. This enables 'allowSyntheticDefaultImports'
      for type compatibility. */

    "forceConsistentCasingInFileNames": true,
      /* Ensure that casing is correct in imports. */

    "strict": true,
      /* Enable all strict type-checking options. */

    "skipLibCheck": true
      /* Skip type checking all .d.ts files. */
  }
}
```

Then, instead of the *.js* file extension, you use a *.ts* file extension. This is configurable and there are many other ways to use TypeScript, but the use of a *.ts* file extension is the simplest.

Now, you write your JavaScript code in a *.ts* file and run it with the tsc command to transpile it into a *.js* file. You then use your *.js* file to bundle and deploy your code. You can optimize this workflow with tools like Webpack.

If TypeScript has issues with your code in the *.ts* file, it'll complain and not generate a *.js* file.

Even without adding static types, TypeScript can infer types from your code and point out any violations. Take a look at Figure 10-1 for an example.

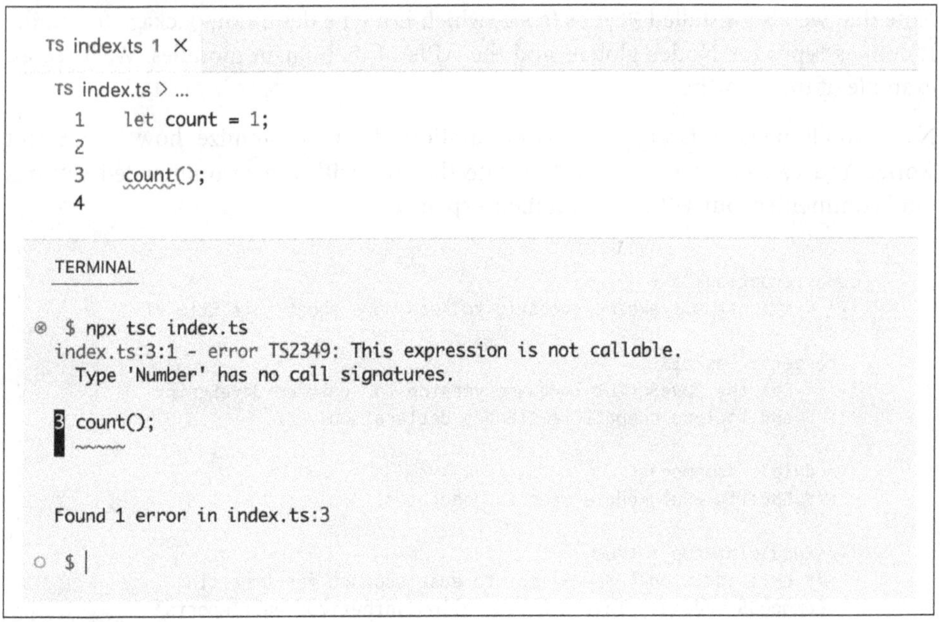

```
TS index.ts 1 ✕

TS index.ts > ...
  1    let count = 1;
  2
  3    count();
  4


  TERMINAL

⊗ $ npx tsc index.ts
  index.ts:3:1 - error TS2349: This expression is not callable.
    Type 'Number' has no call signatures.

  3 count();
    ~~~~~


  Found 1 error in index.ts:3

○ $ |
```

Figure 10-1. A syntax type error

The code declares a count variable and initializes it to 1. Then, it attempts to invoke count as if it were a function. TypeScript blocked that operation. Without TypeScript, since JavaScript is an interpreted language, problems like these might go unnoticed, and the error would only surface when that part of the code was run, in production.

This is an example of an actual TypeError in JavaScript. It's a runtime error that is easy to detect (with many other tools). Logical errors, on the other hand, are more challenging. For example, take a look at the code in Figure 10-2.

This code attempts to increment the count variable. Except, in JavaScript, 1 + "1" is "11"!

TypeScript detected this logical problem without the need of any defined types. This is known in TypeScript as *type inference*. Since count is initialized with a number, TypeScript inferred that type. It also inferred that count + "1" is a string and concluded that this code is trying to store a string value in a variable that has a number type. That's the reported problem.

```
TS index.ts 1 ✕

 TS index.ts > ...
   1    let count = 1;
   2
   3    count = count + '1';
   4


 TERMINAL

⊗  $ npx tsc index.ts
   index.ts:3:1 - error TS2322: Type 'string' is not assignable to type 'number'.

 3 count = count + '1';



   Found 1 error in index.ts:3

○  $ |
```

Figure 10-2. A logical type error

Inference is great but limited, and it does not work in many cases. You should not rely on it. When TypeScript fails to infer a type, it uses the any type (meaning any type can be used). You should avoid using the any type altogether. It exists for a few reasons (for example, to use a third-party library that has no types) but you should completely avoid it in your own codebase.

> You can configure TypeScript (with other code quality tools like ESLint) to make sure no any types are used (both explicitly and implicitly).

Defining explicit TypeScript types is easy; you just add a : *TYPE* to a variable:

```
let isActive: boolean;

let hobbies: string[] = ['Reading', 'Hiking'];

let person: { name: string; age: number };

function add(x: number, y: number): number {
  return x + y
}
```

Note the syntax for defining types for arrays, objects, and functions (arguments + return type). Types can get complicated, but your initial investment in defining them well, from the beginning, will pay off over time.

There are a few things that can make working with types easier. For example, types can be defined once and used many times after. Take a look at the following code and Figure 10-3, for example:

```
type ErrorObject = {
  message: string;
};

function log(error: ErrorObject) {
  console.log(error.message);
}
```

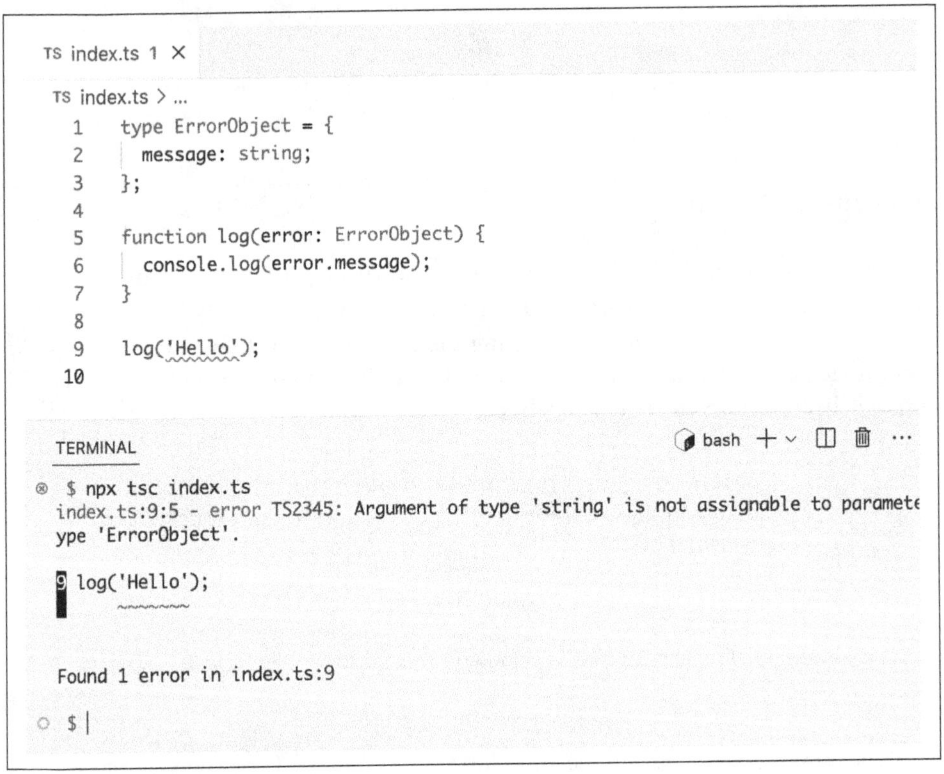

Figure 10-3. Defining types

Types can also be exported/imported and shared with other projects. We installed the @types/node package, which makes the types for all Node modules available in our project, as shown in Figure 10-4.

```
TS index.ts 1 ✕

TS index.ts > ...
    1    import { createServer } from 'node:http';
    2
    3    const server = createServer();
    4
    5    server().listen(3000);
    6


TERMINAL                                              bash  + ∨  ⬚  🗑  …

⊗  $ npx tsc index.ts
   index.ts:5:1 - error TS2349: This expression is not callable.
     Type 'Server<typeof IncomingMessage, typeof ServerResponse>' has no call signature

  5 server().listen(3000);


   Found 1 error in index.ts:5

○  $ |
```

Figure 10-4. Using types from a package

Note how TypeScript detected the problem in `server()`. It knows that `server` has a
type of `Server<typeof IncomingMessage, typeof ServerResponse>` which is not
callable. This is part of the `@types/node` package.

You don't have to worry about defining types for any of the built-in objects in Node.
A huge community of developers has already done that for Node and countless other
libraries (and they continue to update these types as their libraries get updates). For
example, if you need to use the `lodash` package in your TypeScript project, install the
`@types/lodash` to get the TypeScript type definitions for all `lodash` methods. Check
the Definitely Typed project repository (*https://oreil.ly/pA6C3*) to see what libraries
have predefined types.

Some packages even provide their own built-in type definitions. The `graphql` package is an example of that: you don't need `@types/graphql`; once you install `graphql`
you also get all the types of its objects.

Hopefully, this is now the beginning of your journey with TypeScript. I strongly
encourage you to expand and learn more, and use TypeScript in all Node projects and
any JavaScript you run anywhere.

Summary

There are many tools we can use to optimize and improve the development work-flows in Node. These tools make developers more efficient and more productive, and they improve the quality and reliability of the code.

A simple tool like Prettier saves a lot of time and effort by automatically formatting the code consistently. ESLint improves code quality and helps in detecting problems in the code as you type it.

Task runners tools like gulp and Grunt streamline the process of running repetitive tasks in different environments. Module bundlers and JavaScript transpilers help in packaging the code and optimizing it for browsers.

While Node's built-in modules offer great core features for many services, many frameworks evolved to wrap these features, extend them, and make them usable in easier and more efficient ways.

TypeScript extends the JavaScript language itself to offer static types that can be checked at compile time before the code is run. This helps catch errors early during development, effectively blocking developers from writing bad code and helping them come up with higher quality and more reliable code.

Index

Symbols

$ (dollar) sign, 4
* as syntax, 17
--production flag, 15
--save-dev (-D) argument, 13
. (relative path), 49
./ (dot slash), 16
.bat file extension, 150
.catch function, 66
.cjs file extension, 16
.cmd file extension, 150
.js file extension, 15
.mjs file extension, 15
.then function, 66
== operator, 157
?? (nullish) operator, 41
[ ] (square brackets), 36
\ (split into multiple lines), 40
^ (caret) character, 108
~ (tilde) character, 108

A

aborting, 36
Active LTS (long-term support) status, 5
add-on compiled files, 50
AI assistants, 199
algorithms, 177
ApacheBench (AB), 178
Apollo Server, 208
application resources, minimizing, 200
arguments keyword, 51 (see also options and
    arguments)
argv property, 38
Arrange-Act-Assert pattern, 156

Array class, 45
assertion errors, 90
assertions, 155
async function, 27
async generators and iterators, 137-139
async keyword, 63
async/await syntax, 20, 67-69
asynchronous operations (see also event-driven
    model)
    callback pattern, 22
    event emitters, 76-78
    event loop, 73
    import() function, 19
    libuv library, 73
    Promise objects, 23
    sync versus async handling, 59-63
authentication, 191
autocomplete, 45
automation
    automated testing, 97
    loading modules automatically, 39
    in Node, 204
    task runners, 203
availability, increasing, 185-187
availableParallelism function, 180
await keyword, 27, 68

B

Babel, 209
backpressure, 130
.bat file extension, 150
bidirectional streams, 126
brackets, square ([ ]), 36
broadcasting messages, 182-185

buffers, 144

C

caching, 57, 191
call stack, 72
callback hell, 64
callback pattern, 22, 64, 69
caret (^) character, 108
.catch function, 66
chaining, 67, 127
child processes
    benefits of multiple processes, 143
    creating, 143
    execFile function, 150
    fork function, 150-153, 176
    shell syntax and exec function, 147-149
    spawn function, 144-147, 148
    synchronous blocking functions, 150
.cjs file extension, 16
clearInterval function, 20
clearTimeout(timerId) function, 20
cloning, 176, 184
close event, 145
cluster.workers object, 182
.cmd file extension, 150
code editors, 199
code examples, obtaining and using, ix
code quality, 96, 195-200
code reviews, 97
commands (see also Node CLI)
    aliasing, 37
    .break, 44
    cd, 6
    .clear, 44
    echo, 115
    .editor, 44
    exit, 44, 115
    find, 143
    grep, 143
    init, 14
    .load, 45
    ls, 10
    mkdir, 6
    node, 5, 35
    node - C, 36
    node --env-file, 43
    node --eval, 36
    node --import, 38
    node --print, 36

node --require, 38
node --test, 39
node --v8-options | less, 39
node --watch, 39
npm, 4, 30, 102-108
npm -l, 102
npm <command> -h, 102
npm cache clean, 106
npm create, 103
npm help <term>, 102
npm help npm, 102
npm init, 12, 115
npm install, 10, 13, 102, 106
npm install <foo>, 102
npm link, 106
npm list <package>, 106
npm list-unused-packages, 116
npm ls, 110
npm outdated, 110
npm pkg set type=module, 15
npm prune, 112
npm publish, 106
npm run, 115
npm run <foo>, 102
npm run test, 115
npm search <search terms>, 105
npm show, 117
npm show <package>, 104
npm test, 102, 115
npm uninstall, 13, 111
npm update, 108, 109
npm-run-all, 205
npx (node package execute), 117
npx create-react-app, 118
nvm install node, 5
OS commands, 143
ps, 40
pwd command, 144
reload, 193
run npm eslint, 116
.save, 45
wc command, 146
comments and questions, x
CommonJS
    versus ES modules, 8, 15
    exports argument, 55
    implicit arguments in, 52
    module wrapping, 51
    potential drawbacks of, 31

using short absolute paths, 49
compiled files, 50
configurable variables, 53
console object, 6
console.log method, 6
consuming streams, 126, 128
content management systems, 209
context object, 167
continuous integration (CI), 171
continuous integration and deployment services, 204
coverage, 172
CPU-bound tasks, 32
create-react-app package, 118
createReadStream method, 125
createServer function, 9
crypto module, 37
crypto stream, 141
Current release status, 5
custom errors, 89-90

D

Dahl, Ryan, 1, 28
data event, 129, 145
debugging
    Chrome browser DevTools, 94
    fixing bugs, 101
    locating problems, 94
    node inspect script.js, 94
    NODE_DEBUG environment variable, 41
    performance profiler, 95
    SIGUSR1 signal and, 188
    tracing flags, 95
decomposing scalability, 176, 184, 191
deep equality check, 158
default export syntax, 18
dependency injection design pattern, 56
dependency management methods, 8, 11, 54, 112
dependency trees, 100
deployment services, 204
describe method, 158
development frameworks, 205-209
development-only dependencies (devDependencies), 13
directories
    changing active, 6
    creating new, 6
disconnect event, 145

distributed caching, 191
dollar ($) sign, 4
dot slash (./), 16
downloading, 4
downtime, avoiding, 187-190
drain event, 130
dummy objects, 164
Duplex streams, 126, 135
dynamic import expressions, 20
dynamically typed languages, 32

E

EACCES error code, 87
EADDRINUSE error code, 87
ECMAScript standard, 8, 209
ecommerce applications, 209
ECONNRESET error code, 88
editors, 199
emit function, 74
end event, 129
end-to-end (e2e) test, 162
ENOENT error code, 87
ENOTFOUND error code, 88
entry points, 201
env property, 40
environment variables, 40-43, 49
equality, deep versus shallow, 158
error codes, 87
error-first callback style, 64, 92
errors
    assertion errors, 90
    custom errors, 89-90
    error events, 78, 130, 144
    error forwarding method, 92
    EvalError, 87
    input errors, 88
    layered error management, 91-93
    logical type errors, 212
    operational errors, 88
    preventing, 96
    RangeError, 86
    ReferenceError, 86
    standard errors, 85-87
    syntax type errors, 212
    SyntaxError, 86
    system errors, 87
    throwing and catching, 83-85
    timeout errors, 88
    TypeError, 87, 212

types of, 85-91
URIError, 87
user-specified errors, 89
ES modules
accessing filename or directory names of, 53
await keyword, 69
versus CommonJS, 8
implicit scope execution, 52
resolving but not executing, 50
using, 15-19
ESLint package
benefits of, 96, 196
description of, 30
development-only dependencies, 13
using, 198
ETIMEDOUT error code, 88
EvalError, 87
event emitters
asynchrony, 76-78
concept of, 74
creating, 74
error events, 78
examples, 79
main features of, 74
streams, 127, 145
using, 75
event handlers, 3
event loop, 72-74
event queue, 73
event-driven model
adopted by Node.js, 2
benefits of, 59, 75
defined, 3
event emitters, 74-81, 127
event loop, 72-74
handler functions, 63-72
sync versus async handling, 59-63
in V8 JavaScript engine, 1
events
flow of event loops, 74
implicit, 26
registering handlers for, 144-147
slow operations and, 3
stream events, 129-130
exec function, 147-149
execFile function, 150
execute method, 78
exit event, 144, 145
explicit rest parameters, 51

export object syntax, 17
export statements, 15-19
exports argument, 55
Express, 206
extraneous label, 111, 116

F
fake objects, 165
feature testing, 161
FIFO (first-in, first-out), 73
file-compression script, 141
files
compressing, 141
creating large, 123
locating, 16
types supported, 50
finish event, 130
first-in, first-out (FIFO), 73
flowing mode (streams), 130
for loops, 21, 60, 72
fork function, 150-153, 176
fs.readFile method, 63, 124
functional tests, 161
functions
async function, 27
availableParallelism function, 180
.catch function, 66
clearTimeout(timerid) function, 20
createServer function, 9
delaying execution of, 19
emit function, 74
exec function, 147-149
execFile function, 150
fork function, 150-153, 176
ghRepos function, 137
handler functions, 63-72
import() function, 19
import.meta .resolve(), 50
listener function, 9
module.createRequire() function, 51
nested, 72
readFile function, 80
readFileAsArray function, 63, 65
RequestListener function, 9
require function, 8, 15, 38
require.resolve() function, 50
setInterval function, 20
setTimeout function, 19, 22, 61
slowOperation function, 21

spawn function, 144-147, 148
stacked, 72
.then function, 66
timer functions, 23
unref function, 149

G

generator packages, 118
generators and iterators, 137-139
ghRepos function, 137
GitHub Copilot, 199
global scope objects, 6
globalThis property, 6, 53
GraphQL-based API servers, 208
Grunt (task running tool), 203
gulp (task running tool), 203
gzipping, 141

H

handler functions
    analogy for promises, 69-72
    async/await syntax, 67-69
    asynchronous approach example, 63
    child processes and, 144
    error-first callback style, 64, 92
    promises, 65-67
harmony flags, 39
Hello World example, 7
help documentation
    Node CLI, 35
    npm command, 102
    pm2 package, 192
    REPL mode, 44
    tracing flags, 95
Homebrew package manager, 4
horizontal partitioning, 176
horizontal scaling, 175
HTTP/2 protocol, 202
Husky, 205

I

I/O operations, 25
immutable objects, 96
implicit events, 26
import attributes, 50
import statements, 8, 15-19, 41
import() function, 19
import.meta .resolve() function, 50

import/export statements, 52
in-memory state, 191
index.js file, 16
indirect dependencies, 13
inference, 212
input errors, 88
installation (see also npm package manager;
    packages)
    of Node.js, 4
    of packages, 10
integration tests, 162
interprocess communication (IPC), 150
IP hash algorithm, 177, 191
isPrimary flag, 180
it method, 158
iterators and generators, 137-139

J

JavaScript language
    benefits of for Node.js, 3
    list of available globals, 46
    potential drawbacks of, 32
    transpilers for, 209
.js file extension, 15

K

keywords
    arguments keyword, 51
    async keyword, 63
    await keyword, 27, 68, 69
    with keyword, 50

L

last-in, first-out (LIFO), 72
least connections algorithm, 177
libuv library, 73
Linux-based OS, 5
listen method, 9
listener function, 9
listening event, 79, 129-130
Live Server, 205
load balancing, 177
load testing, 178
lodash package, 10, 48
logical type errors, 212

M

machine learning, 200

macOS, 4
Mailchimp Open Commerce, 209
major number, 108
Math.random(), 5
memory consumption, 124, 130
memory management, 39
message event, 145
messages, broadcasting, 182-185
microservices, 176
microtask queue, 74
middleware, 208
minor number, 108
.mjs file extension, 15
mock objects, 165, 166
module bundlers, 200-202
module.createRequire() function, 51
modules
    caching, 57
    crypto module, 37
    definition of term, 99
    descriptions of important, 29
    distinguishing external, 10
    ES modules, 8, 15-19, 50, 52, 69
    executing, 53-57
    full list of, 29, 46
    importing dynamically, 19
    loading, 50
    loading automatically, 39
    managing, 31
    node:assert module, 157
    node:child_process module, 80, 143
    node:cluster module, 80, 176
    node:dgram module, 80
    node:fs module, 24, 37, 123
    node:http module, 125
    node:net module, 80
    node:os module, 80, 180
    node:test module, 97, 156
    node:timers module, 23
    requiring, 38
    resolving, 49
    scoping, 51
    versus scripts, 49
    server.js module, 8, 16
    setting default type, 15
    types of, 50
    using built-in, 7-10, 24, 28-30, 45, 49, 80, 127
multiline mode, 43

multithreaded programming, 2

N

named exports, 17
NestJS, 209
network communication, 126
Node CLI (see also commands)
    creating your own commands, 37
    environment variables, 40-43
    help pages, 35
    node command, 35
    options and arguments, 36-39
    REPL mode, 5, 43-48
Node Version Manager (NVM), 4
Node-API, 50
Node.js
    approach to learning, vii
    asynchronous operations, 19
    benefits of, vii, 1
    executing Node code, 4-6, 36
    fundamentals of, 1-3
    non-blocking model, 21-28, 59, 73
    packages, 30
    potential drawbacks of, 31
    prerequisites to learning, vii
    use of JavaScript language, 3
    using built-in modules, 7-10, 24, 28-30, 45, 49, 80, 127
    using external packages, 10-15
Node.js Toolbox website, 101
node: net module, 80
node:assert module, 157
node:child_process module, 80, 143
node:cluster module, 80, 176
node:dgram module, 80
node:fs module, 24, 37, 123
node:fs/promises module, 27
node:http module, 7-10, 125
node:os module, 80, 180
node:test module, 97, 156
node:timers module, 23
NODE_DEBUG environment variable, 42
NODE_ENV environment variable, 15
node_mksnapshot, 36
node_modules folders, 49
NODE_OPTIONS environment variable, 41
NODE_PATH environment variable, 41, 49
non-blocking model, 21-28, 59, 73

npm package manager, 30, 100 (see also commands; packages)
nullish (??) operator, 41
NVM (Node Version Manager), 4
nvm-windows, 4

O

objectMode flag, 139
on method, 75
operational errors, 88
options and arguments
    error-first argument, 64, 92
    implicit arguments in CommonJS, 52
    overview of, 36-39
OS commands, 143

P

package-lock.json file, 112
package.json files, 11, 101, 103
packages
    creating and publishing, 113-115
    definition of term, 99
    executing initializers, 103
    finding and removing extraneous, 111, 116
    list of, 30
    managing, 31
    npm command, 102-108
    npm loadtest package, 178
    npm package manager, 30, 100
    npm run scripts, 115-117
    npx command, 117
    package management, 99-101
    package registries, 30
    pm2 package, 192
    semantic versioning, 108
    updating and removing, 109-112
    using external, 10-15
parallel processes, 180
passwords, generating, 37
patch number, 108
paths, absolute versus relative, 41, 49
pause() method, 131
paused mode (streams), 130
peerDependencies, 13
performance profiler, 95
performance summaries, 178, 182, 190
piping
    definition of term, 122
    paused and flowing modes, 131

stream pipeline, 127
    to and from child processes, 144
    using pipe method, 125
pm2 package, 192
pressure, of streams, 130
Prettier (code formatting tool), 31, 195-198
primary process, 177-182
Print step, 6
process IDs, 178
process managers, 192
process object, 40
process.argv Array, 38
processes, listing all running, 40 (see also child processes)
--production flag, 15
progress indicators, 141
Promise objects, 69-72
    analogy for promises, 65-67
    versus callback functions, 27
    forwarding errors, 92
    handling asynchronous operations, 20
    purpose of, 23
pull mode, 130
push mode, 130
pyramid of doom, 64

Q

questions and comments, x

R

random algorithm, 177
random method, 10
random strings, generating, 37
RangeError, 86
read() method, 130
Read-Eval-Print-Loop mode, 5, 43-48
readable streams, 123, 127, 130
readFile function, 80
readFileAsArray function, 63, 65, 75
real-time applications, 208
ReferenceError, 86
relative path (.), 49
reload command, 193
repetitive tasks, automating, 203
REPL mode, 5, 43-48
request event, 79
RequestListener function, 9
require function, 8, 15, 38
require method, 11

require.resolve() function, 50
rest parameters, 51
restarting, 185, 187-190
results, printing, 6
resume() method, 131
round-robin algorithm, 177
routing, 205
routing algorithms, 178
run scripts, 115-117
runners, 155-159

S

--save-dev (-D) argument, 13
scaling
    broadcasting messages, 182-185
    definition of term, 175
    handling state, 191
    increasing availability, 185-187
    node:cluster module, 176
    process managers, 192
    strategies of, 175
    zero-downtime restarts, 187-190
scripts (see also commands; Node CLI)
    defining names using pre or post prefixes,
        116, 204
    executing, 36, 43
    file-compression script, 141
    versus modules, 49
    npm run scripts, 115-117
security issues, 32
semantic versioning (SemVer), 108
server.js module, 8, 16
servers (see web servers)
sessions, managing, 191
setInterval function, 20
setTimeout function, 19, 22, 61
shallow equality check, 158
sharding, 176
shell commands, 115
shell syntax, 147-149
SIGUSR2 signal, 188
slowOperation function, 21
snapshots, 36
Socket.IO, 208
spawn event, 145
spawn function, 144-147, 148
special characters, 109
split into multiple lines (\), 40
split method, 37

splitting, 176
spy objects, 165
square brackets ([ ]), 36
stacked functions, 72
standard errors, 85-87
state management, 191
static typing tools, 199
stdio streams, 127
stdout, 6
sticky sessions, 191
Strapi, 209
streams
    analogies for, 121
    async generators and iterators, 137-139
    built-in transform streams, 141
    combining, 122, 141
    data flow in, 123
    definition of term, 121
    fundamentals of, 126
    implementing, 131-136
    major types of, 122
    objectMode flag, 139
    paused and flowing modes, 130
    piping, 122, 127
    stream events, 129-130, 145
    supported by child processes, 144
    using, 123-126
stub objects, 164
synchronous callbacks, 65
syntax checks, 36, 212
SyntaxError, 86
system availability, increasing, 185-187
system errors, 87

T

tab discoverability, 45-48
target audience, vii
task runners, 203
TCP sockets, 126
templates, 118
test-driven development (TDD), 170
testing
    assertions and runners, 155-159
    automated, 97
    benefits of regular, 155
    candidates for, 159
    continuous integration, 171
    example test with context, 168
    failing test example, 159

frameworks for, 199
node --test command, 39
organizing and filtering tests, 169-170
passing test example, 156
selecting tests, 163
test doubles, 163-168
test-driven development, 170
types of, 160-163
with limited resources and connectivity, 200
writing tests, 97
.then function, 66
threads, 2, 59
throttling mode, 200
throw error call, 83, 91
tilde (~) character, 108
timeout errors, 88
timer functions, 19, 23
tools
    automation in Node, 204
    development frameworks, 205-209
    JavaScript transpilers, 209
    module bundlers, 200-202
    quality code tools, 195-200
    task runners, 203
tracing flags, 39, 95
transform streams, 126, 135, 141
transitive dependency, 100
transpilers, 209
try/catch statement, 66, 85
type conversions, 157
type import attribute, 50
type inference, 212
type keys, 15
TypeError, 87, 212
types
    built-in definitions for, 215
    defining, 214
    exporting/importing, 214
TypeScript, 31, 96, 209-215

U

unit tests, 161, 163
unref function, 149
updates
    JavaScript transpilers, 209
    npm update command, 108

for packages, 109-112
trusting npm packages, 101
zero-downtime restarts, 187
URIError, 87
URL hash algorithm, 177
user authentication, 191
user conditions, 36
user-specified errors, 89

V

V8 JavaScript engine
    benefits of event-driven model, 1
    list of V8 options, 39
    starting V8 CPU profiler, 36
versions
    Current versus Active LTS status, 5
    determining, 4
    running multiple, 4
    semantic versioning, 108
vertical scaling, 175
Visual Studio Code (VS Code), 199

W

watch mode, 39
web servers
    creating simple, 7-10
    development frameworks, 205-209
    increasing availability, 185-187
    starting in watch mode, 39
Webpack (asset bundling tool), 31, 201
WebStorm (code editor), 199
weighted round-robin algorithm, 177
window object, 6
Windows, 5
with keyword, 50
words, counting, 37
worker mode, 180
worker processes, 177-182
writable streams, 123, 127
write method, 132

Z

zero-downtime restarts, 187-190
zlib stream, 141

About the Author

Samer Buna has over 20 years of experience in software development, API design, database management, and scalability. He has authored several technical books and online courses about JavaScript, Node.js, React.js, and more. Samer is passionate about everything JavaScript, and he loves exploring new libraries. His favorite technical stack is PostgreSQL, GraphQL, Node, and React.

Colophon

The animal on the cover of *Efficient Node.js* is the red-billed pigeon (*Patagioenas flavirostris*), a large, robust species of pigeon whose primary range extends from Costa Rica through Mexico. They may also be found in southernmost Texas, along a limited stretch of the Rio Grande.

Typical of many species in this region, relatively little is known about the red-billed pigeon. They reportedly eat acorns, seeds, and plant buds that are foraged in the dry forest and riverside wetland habitats where they nest and breed. Their distinctive call has been described as a long, high-pitched "Coooo!" followed by three "Cuk-c'-c'-coo" notes.

The red-billed pigeon has been categorized by the IUCN as being of least concern, from a conservation standpoint. Many of the animals on O'Reilly covers are endangered; all of them are important to the world.

The cover illustration is by Karen Montgomery, based on an antique line engraving from Shaw's *Zoology*. The series design is by Edie Freedman, Ellie Volckhausen, and Karen Montgomery. The cover fonts are Gilroy Semibold and Guardian Sans. The text font is Adobe Minion Pro; the heading font is Adobe Myriad Condensed; and the code font is Dalton Maag's Ubuntu Mono.

O'REILLY®

Learn from experts.
Become one yourself.

60,000+ titles | Live events with experts | Role-based courses
Interactive learning | Certification preparation

**Try the O'Reilly learning platform
free for 10 days.**

9 781098 145194